KARL KRUSZELNICKI

LATEST GREAT MOMENTS IN SCIENCE

ABC
BOOKS

ILLUSTRATED BY KERRIE LESTER

Published by ABC Books for the
AUSTRALIAN BROADCASTING CORPORATION
GPO Box 9994 Sydney NSW 2001

First published 1991
Reprinted August 1991
Reprinted January 1992
Reprinted March 1993
Reprinted May 1995
Reprinted October 1996
Reprinted October 1997

National Library of Australia
Cataloguing-in-Publication entry
Kruszelnicki, Karl, 1948– .
   Latest great moments in science

   ISBN 0 7333 0144 4.

   1. Science. 2. Science—Popular works. I.
Lester, Kerrie. II. Title. III. Title: Great
moments in science (Radio Program)
500

Illustrated by Kerrie Lester
Designed by Geoff Morrison
Set in 11/13pt Plantin by Caxtons Pty Ltd, Adelaide, South Australia
Printed and bound in Australia by Australian Print Group, Maryborough, Victoria

Contents

Dedications

Chris Norris, who edited the scripts, made them suitable for radio station JJJ and added music,

Graham and Annie Beatty, who gave punchlines,

Mary Dobbie, who listened and criticised,

and David Malin, the only man I have ever met who has an entire galaxy named after him, and who was kind enough to criticise the astronomical stories.

Animals Get High

i t's not just humans that get high, animals and birds do it all the time! But why do we do it?

Ducks love to get high on all the narcotic plants, including the opium poppy. Jaguars in Columbia chew the hallucinogenic bark of the yaje plant. In West Africa, wild boars dig up and eat the hallucinogenic roots of iboga, a shrub with pretty flowers. And in Hawaii, dogs and cattle go for the magic mushrooms, which are loaded with psilocybin, another hallucinogenic drug.

Siberian reindeer love eating the red-capped hallucinogenic mushroom *Amanita muscarita*. The Siberian Shamans or witch-doctors also eat the sacred mushroom. Only a little bit of this drug is used up in the body; most of it passes out with the urine. The Siberian Shamans drink their own urine so they can have a second go at their hallucinogenic drug. That's really getting pissed. But the reindeer know this trick too. In fact, travellers in that part of Siberia are warned not to urinate in the open when the reindeer are around. Rudolf the red-nosed reindeer would try to drink the urine before it hit the ground.

African elephants are more traditional. They go for rotting fruit. Rotting fruit can naturally ferment by itself up to 7 per cent alcohol. That's 2 per cent more alcoholic than beer. In 1974, 150 elephants drank litres of high-grade alcohol in an illegal distillery in West Bengal. By the time they smashed through seven concrete buildings and 20 village huts, they'd injured seven humans and killed five.

In Bundaberg, the rum capital of Queensland, the local lorikeets get smashed on spilt sugar, which falls off trucks making deliveries to the rum factory. At night, dew settles on the road, and dissolves the sugar. The rays of the sun help to speed up the fermentation into alcohol, and the pickled parrots then get wingless, and get run over by the road traffic.

Pigs love marijuana. In fact so many animals love marijuana that they've caused a major socio-economic-scientific change in California—they've stimulated the alternative energy market. In California, the humans who buy the most solar cells are the growers of illegal marijuana. They use the solar cells to power up electric fences. The electric fences keep the drug-hungry animals away.

Legend from AD 900 has it that an Abyssinian herder discovered coffee. He noticed that his animals became very energetic after eating the bright red fruit of the coffee tree. The amphetamine stimulant called quat was discovered by a shepherd in Yemen. He noticed that his goats became hyperactive after they chewed on the leaves.

The animals have really taught us a lot of bad habits. An ancient ceramic bowl from the Peruvian Andes has a painting of some llamas and some Indians. The llamas are eating some coca leaves—it reduces their thirst and helps them to work better at the high altitudes—and that's where you get cocaine from! The Indians are pointing at the llamas and are reaching for the coca leaves with both their hands and mouths open.

But birds from 5000 years ago gave us the first blood-pressure drugs! In the Vedas, ancient Hindu scriptures that date back 5000 years, there are some medical textbooks. One of them mentions the rauwolfia shrub. The ancient Hindus used extracts from this shrub to treat high blood pressure and insanity. Their legends say that originally they had noticed that birds became strangely quiet after nibbling on the shrub. Following up on these legends, in 1931 Indian scientists isolated a drug from the roots. They called it reserpine. It was one of the first drugs used to treat high blood pressure and insanity.

Drugs are at all levels of society. In a 1951 cartoon film, Donald Duck drank some peyote tea—he went off his tree and was in a coma for six weeks. Dumbo the elephant, Wile E. Coyote, Fritz the Cat, Porky Pig, Woody Woodpecker and even Mr Health Food, Bugs-carrot-Bunny, have all gotten their rocks off with drugs ranging from alcohol to marijuana and the hallucinogens. And even in the soapies, the rich and famous can't talk without a glass of alcohol in their hands.

But most of these drugs are a bit poisonous. They often taste bitter, cause nausea and vomiting, interfere with the activity of the liver, or even cause death and

mutations in the next generation. So why do animals and humans take drugs?

One excuse is to relieve fatigue. The tired horses in the rugged mountains of Sikkim, in north-eastern India, love tea leaves because of the stimulant effect of the caffeine. And when pack donkeys in Mexico are being over-worked, they seek out wild tabacco. They don't seem to mind twitching and trembling and having diarrhoea, so long as they get that tobacco high and can keep on working.

Another excuse for taking drugs is to relieve psychological stress. Under normal conditions, when they have lots of room, elephants are quite happy to sip small quan-tities of alcohol from their rotting fruit. But when they were squashed into a smaller living area, they guzzled it down. Just like humans, some were aggressive, some sleepy and others became horny. When they had more living area, they settled back down again.

But some people get addicted accidentally. Just look at Marty Mouse. Marty was a small mouse that accidentally got locked into a vault held by the San Jose Police. In the vault were kilograms of marijuana, amphetamines, narocotics—every drug that they had ever busted. He had nothing to eat but unlimited quantities of the very best street drugs. He had no choice—it was either drugs or starve to death.

But like all big druggies, Marty got busted. Ron Segall from the local university took Marty back to his lab for experiments. He ignored the protests of the Free Marty Fan Club. He found that not only Marty, but all rats in the lab ate marijuana seeds because they are low in those nasty cannobinoid drugs but high in nutritious edible oils. But the leaves have more essence of marijuana. While the other rats wouldn't eat leaves, Marty would—he was really addicted. First he would become quiet and later aggressive. He slept more, but on the other hand he spent

more time making love to his wife, Mary Jane Mouse.

But in our society, we usually get addicted to tobacco and alcohol because of pressure to be the same as our friends—because we don't like to be different.

So here's the secret to dealing with drugs. Do without. And if you have to take drugs (and most animals seem to) have as little as possible. And take it in its natural form.

Take, for example, coca. The coca leaf has less than 1 per cent cocaine. The rest is fats, carbohydrates, proteins, minerals and vitamins. This makes the coca leaf a nutritious pick-me-up for the animals and the Peruvian Indians. And because it is so diluted, it's not very dangerous. No animal or Indian has ever died from eating coca leaf. But when it's purified to cocaine, it's easy to overdose and have a bad trip, or even die.

Your parents were right—moderation in all things.

REFERENCES

Omni March 1986, pp 70–74, 100
The Pharmacological Basis of Therapeutics Goodman and Gillman 6th edition, Macmillan, New York, 1980, pp 202–204

Anti-Anti-Anti-Tank Weapons

*W*henever the military want to take a city, whether it's Beijing, Panama or Beirut, they don't call in the airforce or the navy—they send for the tanks. There are more than 160 000 tanks in over 100 countries. There aren't many ways that a civilian can stand up to 65 tonnes of rumbling steel monster.

Ever since they were invented in 1917, tanks have been the ultimate fighting vehicle. They were so frightening that in Word War I, there was a new disease on the battlefields—tank fever. The infantry would drop their weapons and run at the sight of them.

Since then, weapons scientists have been searching for the ultimate anti-tank weapon. By World War II there was the bazooka—a rocket launched from a hollow tube called a rocket-launcher. It was light enough to be carried and fired by a single soldier, but was not very accurate. The balance of power still lay with the tank.

But in the Arab–Israeli Yom Kippur war of 1973, Egyptian soldiers were given Soviet wire-guided 'Sagger' missiles to play with. The soldiers fired the rocket in the general direction of the tanks, and then 'steered' it with a joystick, just like a real live video game. This was the first time that the foot soldier had overwhelmingly defeated tanks, and afterwards hundreds of kilometres of wire strands lay across the desert, like a huge collapsed spiderweb.

So the tank scientists went back to the drawing board and invented a new type of armour—Chobham armour. It had many layers of steel and ceramic—ceramics are harder than metal—and stopped conventional projectiles.

But even with new armour, tank supremacy lasted only five years. By the late 70s, the anti-tank scientists invented a fantastic new weapon—powerful enough to blow a hole through 800 millimetres of armoured multi-layered steel, but light enough to be carried by a single soldier.

This anti-tank weapon is called HEAT, standing for high explosive anti-tank. It has a small special warhead inside a pointed missile roughly the diameter of your fist and as long as your arm, and is fired from a rocket-launcher. The warhead is a cylinder of high explosive, about the size of a fat salami. This cylinder of explosive has a hole, shaped like a hollow cone and lined with metal, at the front end. When the charge goes off, the thin metal cone collapses into a skinny long jet of red-hot metal, the shape of a knitting needle, but moving at between 10 and 50 kilometres per second. When this hypersonic knitting needle touches the armour plate of the tank, it pushes a hole through the armour steel, like a jet of water from a hose can push a hole through a sand castle. The crew inside the tank are killed, not by heat or fire, but by the enormous pressure as the jet bursts through the armour.

But this HEAT missile had the upper hand for only a short time, because the tank scientists foiled the poor foot soldier with armour-that-goes-bang-and-fights-back. It's called reactive armour and is simply blocks of explosive that are bolted onto the outside of a tank. Each block is sandwiched between two steel plates and is roughly the size of a telephone book. It reacts to the incredibly fast jet of metal from a HEAT shell by exploding outwards at the moment of impact. Its own outward explosion weakens the destructive effect of the incoming knitting needle by about 75 per cent. Reactive armour adds about a tonne to the weight of an M60 tank—but it gives the same protection as 10 tonnes of steel. Having the armour in separate blocks ensures that they won't set each other off when they explode.

But reactive armour can kill your own troops. When it explodes, the front steel plate is blown away from the tank at enormous speed. The Israelis invented it, and when they first used it in 1982 in the Bekaa Valley campaign during the Lebanon War, several Israeli soldiers were killed by flying steel plates from the reactive armour.

But even worse for military secrecy, a few M60 tanks that were equipped with reactive armour were captured by Syrian forces and sent back to the USSR. So by 1986, reactive armour was appearing as a retrofit on the Soviet T64 tanks.

Reactive armour is an anti-anti-weapon that protects the tank by going bang once. So the anti-tank scientists came up with an anti-anti-anti-weapon to attack the tank—a missile-that-goes-bang-twice. Basically you have two weapons on the same missile, one behind the other. The first weapon gets rid of the reactive armour to allow a free path for the second weapon. And already you can buy off-the-shelf on the military marketplace two different anti-anti-anti-weapons.

So the seesaw swung in favour of the anti-tank team again. The tank scientists had to come up with a counter weapon for the-missile-that-goes-bang-twice. Of course they did; they invented an anti-anti-anti-anti-weapon to protect the tank. It was armour-that-goes-bang-twice.

Some Soviet tanks now have multiple layers of reactive armour that will stop anti-tank rockets with dual weaponheads. So the Soviets have taken the quick and easy way, with double armour.

The Americans have gone down a slightly different path—armour-that-goes-bang-once plus funny-metal. They've come up with depleted uranium, that can be used with the tile of reactive armour.

It's called depleted because it has practically all the atom-bomb-grade uranium taken out of it. As a metal uranium is very soft and ductile—not at all suitable for anti-anti-anti-anti-tank weapon armour. The word is that the Americans have done some very fancy metallurgy, and have made the armour plate out of a whole bunch of monocrystals of a uranium compound. Monocrystals of anything are very strong. A monocrystal turbine

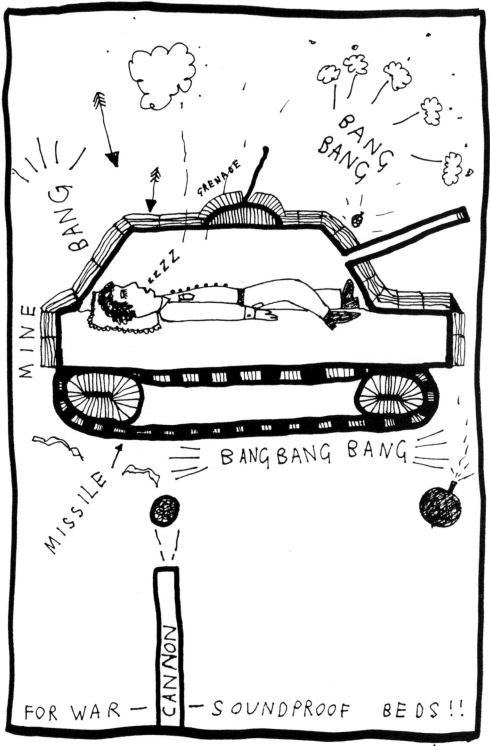

fan blade, when it's used in a jet engine, is roughly the size of the palm of your hand and can extract 500 horsepower from the moving air stream.

This depleted uranium is to be fitted to 3000 of the 4500 M1 Abrams Main Battle Tanks. Its radiation is claimed to be less than the normal background level. But on the other hand it does release poisonous radioactive gases, probably radon, when it breaks up—but the crew of a tank whose armour is disintegrating will have more to worry about than poisonous gases.

In a nuclear blast, the uranium armour will react with the neutrons released from the nuclear weapon, and will triple the number of neutrons inside the tank—so the crew will be actually worse off if they are inside the tank than if they were just sitting under a tree. But somebody has to pay the price of military progress.

Now you'd think that this ridiculous see-sawing game of weapon versus anti-weapon and armour versus anti-armour had gone far enough, if you have to protect yourself by carrying around both explosives and uranium. But Marconi went one step further. They've come up with a tank anti-missile system that has two radar-controlled 7.62 millimetre chain guns. They carry 100 rounds of ammunition each, and will fire 20 rounds of ammunition in an encounter with an incoming missile travelling at a kilometre or so per second. So this is moving from passive defence to active defence—or is that aggression?

You might think that all of this to-ing and fro-ing would have led to some sort of stable balance. But it hasn't, because these super-sophisticated anti-tank weapons cost about $50000 each. They are too expensive for the infantrymen to practise on. In fact they've spent so much money developing these weapons, they can't afford to buy them. So the Soviet Union has more tanks than the Americans have anti-tank weapons.

But just to show how ridiculous is this waste of money, the British have come up with a brand-new anti-tank weapon—the statistical mine. It's a mine that can be dropped onto a battlefield from an aeroplane and will then move around in a completely random and unpredictable manner. Not even the people who put it there know where it will end up.

Four of every five physicists work for the military on projects like these. Seventy-five years of well-funded research have gone by, and the balance of power between the foot soldier and the tank has not really changed since World War I. Maybe that Chinese civilian in Beijing uprising of 1989 had the right idea when he stopped the tanks with two shopping bags.

REFERENCES

Popular Science April 1988 'Tank Armour that Fights Back' by Stuart F. Brown, pp 62–63, 114
Military Technology October 1988 'The "Hows" and "Whys" of Armour Penetration' by Professor Giorgio Ferrari, pp 81–96
Sydney Morning Herald 22 July 1989, p 24
New Scientist No 1686, 14 October 1989, p 19

Ball Lightning

*b*alls of lightning have been invading highflying jets and terrorising the passengers.

Ball lightning is a really weird type of slow-moving lightning that moves through walls and likes enclosed spaces. It's usually in the shape of a sphere, but sometimes looks like a rod or a dumbbell. It is usually 15 centimetres in diameter, but it can be as small as a pea, or as big as a car.

Ball lightning can float casually and silently around the landscape at walking speeds, or it can nose around slowly into all the corners of a room and even enter tightly sealed rooms. It lasts for up to five minutes and usually happens just after a lightning stroke. It can be any colour, but purple and green are not very common. It usually shines steadily, but sometimes it pulsates. And it can vanish silently, or with a big bang.

Now the first recorded case of ball lightning was in 1754. Dr Richmann of St Petersburg in Russia tried to duplicate the famous kite–lightning experiment of Benjamin Franklin. Lightning struck the lightning rod, and a glowing blue-white sphere of ball lightning, the size of a baseball, erupted from his measuring apparatus, and struck him in the head, killing him instantly. The only evidence left behind was a red spot on his forehead and two small burn holes on one of his shoes.

In London in 1936, a man saw an apple-sized ball of lightning land in a bucket holding 15 litres of water. The water actually boiled for a few minutes, and it was still hot enough to scald 20 minutes later. Scientists at the time refused to believe this report, as well as the thousands of others on file.

But this all changed five minutes after midnight on a stormy night, 19 March 1963. A bunch of scientists was travelling on an Eastern Airlines flight over New York City towards Washington. Roger Jennison, professor of electronics at Kent University in the UK looked up to see a glowing sphere about 20 centimetres in diameter come out from the wall of the pilots' cabin and pass down the aisle of the aircraft. It looked solid and was blue-white. It drifted through the cabin less than a metre above the floor, at a normal walking pace. Then it merged with the skin of the aircraft at the rear and passed out into the night. Suddenly, ball lightning became respectable.

There's a Soviet report of a 10-centimetre ball of fire invading an Ilyushin-18 aircraft on 15 January 1984. It suddenly appeared on the outside of the aeroplane in front of the cockpit as the plane flew close to a thunderstorm. It then disappeared with a deafening noise, but suddenly reappeared several seconds later in the passenger compartment. It slowly drifted down through the cabin. Then near the tail of the aircraft it divided

15

into two segments which then recombined into a single ball. It then merged with the metal skin of the plane and also vanished into the night, leaving behind a wrecked radar unit and several dead radios.

No one can yet explain how ball lightning happens, but scientists in Holland are trying to make an artifical ball of lightning, to get cheap electricity from fusion. Fusion is basically a slowed-down hydrogen bomb. The colossal heat of fusion will melt any solid container. So most of the fusion research of the last 30 years has been focused on making a magnetic bottle, an invisible force field, to contain this super-hot reaction. They *can* make a magnetic bottle, but it lasts for only a few squillionths of a second.

The Dutch scientists think that the ball in the ball lightning would be a good container for a slowed-down mini hydrogen bomb. If they're correct, cheap and fairly clean electricity, inside a ball of lightning, could be just around the corner.

REFERENCES

Scientific American March 1963 'Ball Lightning' by Harold H. Lewis, p 105
New Scientist No 1592/1593, 24/31 December 1987, pp 64–67

Bonfire is Bad

*a*cross Australia, more and more councils are making it illegal to burn off rubbish in your backyard. It may seem an infringement of your civil rights, but a backyard bonfire can do more damage than chain smoking.

We get the modern word bonfire from 'bone-fire'—the fires made in medieval times from the bones of oxen and sheep. In Olde Englande, they lit these smokey fires to ward off evil spirits, and to keep warm.

Now you'd think that if you burnt ordinary organic garden rubbish like grass clippings, twigs and branches, you'd get carbon dioxide, water vapour, and a small amount of mixed organic gases (various oxides of nitrogen from the nitrogen in the DNA and the proteins). But this clean burning happens only in laboratories, when you have a super-efficient hot fire, with a lot of oxygen fed to it. In the typical backyard burning heap it's only the outside of the rubbish pile that gets enough oxygen to make carbon dioxide; the inside of the bonfire makes carbon monoxide.

Carbon monoxide is that famous, but unnamed, gas in the suicide scenes in movies, where they pipe the exhaust gases into the cabin. When you breath it into your lungs, it sticks onto blood cells about 250 times better than oxygen does, so your blood cells can't carry enough oxygen. Your average 1-tonne bonfire will give off about 30 kilograms of carbon monoxide. In fact, one fire can pollute your suburban street with so much carbon monoxide that your air will be as dirty as downtown Tokyo.

The other problem with a bonfire is that the garden clippings are usually moist. A wet bonfire means unburnt particles floating around, and chemical reactions that make killer hydrocarbons. You see the unburnt particles as clouds of dark smoke. Our 1-tonne bonfire will give about 9 kilograms of incompletely burnt solids. About a third of the brown haze you see over our capital cities comes from backyard bonfires. Some of these unburnt particles are small enough to miss being caught by your natural filtering mechanisms, and can lodge deep inside your lungs.

The killer hydrocarbons include irritating compounds like acetic acid or vinegar and dangerous cancer-causing chemicals like the benzopyrenes. In fact, an English study in the 1950s showed that bonfire smoke has 350 times the level of benzopyrenes as cigarette smoke. And you don't have to breathe in the concentrated bonfire smoke either—research into passive smoking has shown that you can get cancer even if the chemicals are very

diluted. If you have a tendency to suffer from asthma, lung infections or chronic bronchitis, a bonfire can be just the thing to topple you off your perch. Children suffer a much greater risk than adults, because they breathe in more air per kilogram of body weight and because they have a different lung structure. A child sucks six times as much nasty chemicals into its body as an adult.

Burning natural organic garden rubbish is bad enough, but it gets much worse if you chuck in some modern synthetic rubbish. Some nasty chemicals contain chlorine, such as polyvinyl chloride (PVC), which is found in the insulation on electrical cables, synthetic leather, vinyl floor tiles or any plastic coated fabrics. If you burn these, your bonfire will give off the corrosive gas hydrogen chloride. But if the fire is not burning hot enough, and is running at less than 1100°C, your unthinking neighbour's innocent-looking bonfire will give off dioxins, which are the active ingredients of the famous herbicide agent orange. You'd also get phosgene which was the deadly poison gas used in the trenches in World War I. There are 75 other really nasty toxins that are so far known to come smoking out of chlorine-based rubbish.

You get a different bunch of chemical nasties if you burn synthetic materials that contain nitrogen. Nitrogen is in nylon and also in polyurethane foams which are found in mattresses, sofas, arm chairs and some foam-backed carpets. At temperatures above 600°C, these nitrogen-nasties give off hydrogen cyanide. And if you try to burn them when they're moist, and get a bonfire running below 600°C, they give off a thick choking yellow smoke full of isocyanates. Isocyanates became famous in the terrible Bhopal disaster in India. So hot or cold, nitrogen is nasty.

Even burning odds and ends of wood that are left over from doing-it-yourself can be dangerous. Chipboard gives off the cancer-causing formaldehyde. Melamine, that plastic-covered wood found in kitchens, gives off formaldehyde if there's a lot of oxygen, and hydrogen cyanide if there's not. Even old wood from demolished buildings is not safe. They used to use pentachlorophenol as a preservative, and when you burn it at low temperature you get—you guessed it—dioxins. And very old wood is almost guaranteed to have lead in the paint.

You've probably gone for a drive in the countryside in winter, and have seen how the smoke from one single chimney can pollute a whole valley. Imagine what it's like in the city, when half the houses in a street are burning wood. Winter mountain resorts in the USA, such as Aspen and Vail in Colorado, now have prohibitions on woodstoves, because the previously super-clean mountain air was thick with smoke.

So before you go to burn off, think of what you might be breathing in. Also, check with your local council to see what the situation is—you could be hit with a heavy fine. And if you must light a fire outdoors, make it a barby, and use charcoal or clean dry wood—otherwise you'll end up breathing and eating a nasty chemical cocktail.

REFERENCES

New Scientist No 1637, 5 November 1988, pp 48–51

Coal-Powered Laser Rocket

*a*merican military scientists are working on a rocket powered by a laser beam and coal—that's coal, the stuff that you dig out of the ground. But it won't be like a steam train. It's like something out of a Jules Verne novel, a coal-powered rocket!

As part of the Strategic Defence Initiative, or Star Wars, the United States Army is building the world's largest laser. It's called the Ground Based Free Electron Laser, and they are building it near Oro Grande in New Mexico. They are going to use this laser to push cargo into orbit.

The whole rocket is just the payload, or cargo, and a lump of coal—nothing else. The plan is to have the payload on top of a block of coal. Then they will blast a super-powerful laser beam onto the bottom of the coal. The laser heats up the coal, and a thin bottom layer of its surface vaporises and moves down away from the coal. Newton's Third Law says that 'for every action there is an opposite and equal reaction'. So as the gas is pushed down the payload is pushed up.

In fact, they are thinking of kicking this lump of coal into orbit with double pulses of laser energy. The first pulse of laser energy would be fairly weak. It would create the gas. The second pulse of laser energy would be much stronger. It would heat up this film of gas and make it expand very rapidly. This would push the lump of coal, with its payload, up into the sky.

But what's so good about coal? Coal is rich in hydrogen, and of all the gases in the entire universe, hydrogen is the gas that moves the fastest. The faster the gas moves down, the faster the payload moves up. Materials that are rich in hydrogen are ice—that's right, ordinary frozen water—blocks of plastic, and even the coal that you dig out of the ground.

But why go to all the bother to launch a cargo into space with a laser and a lump of coal, if we already have rockets that work quite well? The answer is that coal rockets can lift more payload. The trouble with ordinary rockets is that as well as carrying the payload, they have to carry a whole lot of other rubbish—fuel, complex pumps and plumbing, strong casings and nozzles. In a space shuttle, only 1.5 per cent of the actual lift-off weight is payload. But with the coal—

WHEN COAL WAS A BABY HE DREAMED OF THE STARS AND THE MOON

SO WHEN HE GREW UP HE WENT TO ASTRONAUT SCHOOL

MEANWHILE: LASER, A BEAUTIFUL RED HEAD ALSO AT ASTRONAUT SCHOOL TOOK ONE LOOK AT COAL AND DECIDED THAT HE WAS THE MAN FOR HER. A MATCH MADE FOR HEAVEN

PULSE PULSE PULSE

LASER

GROUND BASED FREE ELECTRON

STEAM

THINGS STARTED GETTING VERY HEATED UP. COAL AND LASER DECIDED THEY MUST BE TOGETHER ALWAYS. THEY WERE MORE THAN JUST AN ITEM. THEN AS IF BY MAGIC WITH ALL THE PULSING AND STEAM THEY TOOK OFF TO THE HEAVENS.........

TOGETHER FOR iEVER!!

COAL LOVES LASER

Lester

laser rocket, 15 per cent of the total lift-off weight would end up in orbit. That's 10 times better.

The US Army hopes to have the first tests in 1991. They will just bolt some coal to a test stand and blast laser beams at it to see what happens. If these tests come out OK, they would then blast laser beams at blocks of coal tied to the ground with ropes. These tests are planned for 1991 and 1992. Gradually they will work their way up to a launch with a block of material that is not tied to the ground. And it's easy to aim the payload. If you want to send your payload to the east, you simply blast the laser beam onto the west side of the coal.

But this current model of the free electron laser is only powerful enough to deliver something the size of a grapefruit to a space station at an altitude of 400 kilometres. Unfortunately, it sounds suspiciously like a weapon. It could be used to launch flak to destroy enemy satellites, just like anti-aircraft guns in World War II.

Now even though the laser itself would cost many millions of dollars to build, the electricity to run it would be very cheap. It would cost only $18 worth of electricity to launch a 1-litre carton of milk into orbit, compared to a price tag of $10 000 for the space shuttle. Maybe this would be the way to send up the lumps of steel to be used in building a space station.

So in the early twenty-first century, the astronauts might really say 'Beam me up, Scotty'. And Scotty would say 'One lump or two', depending on how far they want to go.

REFERENCES

IEEE Spectrum September 1987, p 19

Cold Iron-Deficient Women

*i*f you're feeling the chills this winter, you might need to nibble some real iron-person food. But it's *not* a sugary breakfast cereal, it's caviar. It's all to do with iron.

The average grown-up human body has only 3.5 milligrams of iron and without it, you'd suffocate. If you melted all that iron down, you'd end up with a tiny ball bearing. But there's no pure iron metal floating around free inside your body, because it's very toxic. Instead, iron is stored as a sort of organic rust. Most of the iron is in your blood. Rusty iron is red on your car, and it makes your blood red as well. Your red blood cells pick up oxygen from your lungs and deliver it through the blood vessels to wherever it's needed, like an exercising muscle, or your exercising brain. The iron in the red cells sucks up oxygen from the lungs like a magnet.

The average person eats about 15 milligrams of iron daily. But only about 1 milligram is actually absorbed—the rest, about 95 per cent of it, just passes out your backside into the sewerage system. Now if you're a man, that 1 milligram a day exactly balances what you lose. But women have an extra pathway for losing iron. They lose about

35 millilitres of blood in each menstrual period. So on average, women lose twice as much iron as blokes, but they make up for it by soaking up more iron from their food.

When people don't get enough iron, they get weak and tired, lazy and lightheaded. It's a bit like sticking a sock into the air inlet of a car engine. Too little iron means you can't make enough red blood cells, and about 2 per cent of adult men are iron-deficient. But for women of child-bearing age, the percentage is 10 times higher—20 per cent! One in every five women is breathing through a sock; she's not getting enough oxygen in.

And when a woman becomes pregnant, it gets worse. A growing baby will steal about 15 per cent of the mother's iron, even if she's already low. And if women don't do anything about this, more than half of them will be seriously iron-deficient when they give birth. About 10 per cent of two-year-old children are iron-deficient. There's just not enough iron in milk and ordinary cereals—they need meat

and vegetables and iron-added baby cereals.

But even if you do the right thing and eat iron-person food, you won't automatically get better. Your body has to absorb it, and gluey stuff like organic iron is hard to move across the wall of your gut and into the blood stream. Some food preservatives will actually stick onto the iron while it's in your gut, so that it can't enter your blood stream. Iron sticks to chemicals like the tannin in tea, and phytates in wheat, so a tea-and-toast diet is a quick way of flushing the iron that you do eat down the toilet. On the other hand, vitamin C will help you absorb iron into your blood.

Henry Lukaski, a very cruel researcher with the US Department of Agriculture, wanted to find out just what a low-iron diet did to women. So he fed six volunteers a diet low in iron for six months. He then sat them in a cool room at 18°C, while they wore bathing suits. They were allowed to leave the room only when they began to shiver. Then the women were fed a diet rich in iron for 100 days, and put into the same cold room again. He found that women fed iron-rich food withstood the cold for an extra eight minutes before shivering. And even though they were in this cold room for eight minutes longer, their body temperature fell only half as much.

John Beard, a professor of nutrition at Pennsylvania State University did a more rigorous experiment on 26 women with freezing cold baths, and the results were the same.

Both researchers also found elevated levels of the stress hormone, noradrenalin, in the iron-deficient women. This is probably an attempt by the sympathetic nervous system to compensate, but it could lead to a long-term general 'stress' problem.

So if you're low in iron, you'll feel the cold sooner and you'll feel it more. This winter you can fight the shivers with food. Meats like liver and kidney are rich in iron, as are seafoods like oysters, sardines and, if you can afford it, caviar. Beans like soya beans and lima beans have a reasonable amount of iron, but strangely, practically all vegetables are low—even spinach. Popeye was wrong. The actual discovery that spinach had as much iron as meat was made way back in the 1890s. During World War II, American Popeye propaganda cartoons encouraged the populace to eat lots of spinach—a good thing when there was not much meat around. In fact people ate 35 per cent more spinach, and the people of Crystal City in Texas put up a statue to Popeye to commemorate that famous 35 per cent. But it's all wrong. The original scientists way back in the 1890s did their experiment right, but they wrote the result down wrongly. They put the decimal point in the wrong place. They overestimated the amount the iron in spinach by 10 times. To get his iron, Popeye would have been better off chewing on the can.

If you're a big meat-eater, it's easy to fight the cold with a steel woolly jumper on the inside. But if you want to be a vegetarian iron-person, you'll be eating parsley and beans all winter.

REFERENCES

British Medial Journal Vol 283, 19 December 1981 'Fake' by T. J. Hambin, p 1671
Metric Tables of Composition of Australian Foods by Sucy Thomas and Margaret Corden, AGPS, Canberra, 1977, pp 6–26
Harrison's Principles of Internal Medicine 11th edition, McGraw-Hill, USA, 1987, pp 1489–1498
JAMA Vol 260, No 5, 5 August 1988, p 607
Science Digest March 1989, pp 50–51

Comets made the Oceans

astronomers have been seeing spots on the sun for centuries. Now, with satellites, we can look back at our own fragile little planet and discover dark spots in its upper atmosphere.

Louis Frank, a physicist from the University of Iowa thinks that these black spots are the splashes of comets hitting our atmosphere. His photos show dark spots in the even glow of ultraviolet light that wraps around our planet. The spots are about 150 kilometres in diameter. In these spots, which happen about 20 times per minute over the whole planet, and survive for a few minutes, the ultraviolet light drops to about 10 per cent of its normal level. And what absorbs ultraviolet light? Water.

So Louis Frank has a theory that the water in our oceans came from outer space, carried here by comets—small cosmic buckets of frozen water—about 4.5 billion years ago, when the solar system had just formed. In those days there were many times more comets and meteors than there are today. We know this because of the many craters on the moon. There were more craters on the earth, but the earth is still geologically active, so there is very little evidence of the craters. They reckon that about 10 million billion tonnes of stuff hit our planet. If only 10 per

cent of it were comets (1 million billion tonnes), and 50 per cent of the weight of the comets was water (500 thousand billion tonnes), this would have been enough water to supply about half the water in the ocean. This is completely different from the old theory, which says that the water was cooked out of the molten rock and discharged out of volcanoes soon after the planet formed billions of years ago. Frank's theory is that the earth is continually being bombarded by 20 small comets each minute. That's about one comet every three seconds. The comets are moving at less than 20 kilometres per second, are each about 12 metres across, and weigh about 100 tonnes. They are basically dirty snowballs—a mixture of dirt and ice.

As each small comet approaches within about 3000 kilometres of the earth, it breaks up. At that altitude, there is not much atmosphere, so the comet spreads out, from a small 12-metre iceball, into a cloud of steam 50 to 150 kilometres across. This piston of steam plunges into the upper atmosphere, and absorbs the ultraviolet light. They're the

black spots that the satellite has been seeing.

These comets dump about one billion tonnes of water onto our planet per year. This is enough water to raise the level of the ocean by 2.5 centimetres in 10 000 years. But of course, some of the water evaporates back into space again.

A watcher on the ground would see a thin streak of light—they are called 'Opik's Dustballs'—which would last about one second. It would be about as bright as Venus, and you would expect to see one every 100 hours of watching.

Of course, other scientists were sceptical so they photographed the sky with a very sensitive electronic camera. To their surprise, they saw faint streaks, of exactly the predicted length and brightness.

One billion tonnes might seem like a lot of water, but it's only a drop in the ocean compared to the greenhouse effect.

REFERENCES

Scientific American July 1986, pp 56–57
Geophysical Research Letters No 4, Vol 13, 1986, pp 303–310
Geophysical Research Letters Vol 13, No 11, November 1986, pp 1181–1183
New Scientiest No 1612, 12 May 1988, p 38

Cycling Supercontinents

Continents come and go, but if you want something really permanent, go for the Pacific Ocean. All the other oceans are just temporary.

Our planet is about 4.5 billion years old. In the last 2.5 billion years all the continents have collided six times to make one giant supercontinent. That supercontinent lasted 80 million years, before it busted apart into smaller continents like those we know today. This cycle of supercontinents and continents has controlled the climate on our planet, and has ultimately chosen what sort of life we see here on earth.

The most recent supercontinent is called Pangaea. It formed about 300 million years ago, and it stretched between the North and South Poles. It was the only land on the planet, and it was surrounded by a single huge ocean, the Pacific Ocean. Today the Pacific Ocean is much smaller than it used to be, but soon, in only a few hundred million years, it'll cover more than half the earth again.

It was heat from the decay of natural radio-active elements inside the earth that broke up Pangaea. A lump of land is like an industrial-grade doona—a good insulator. The crust under the continents is thicker than the crust under the oceans, and it traps more heat. Gradually, the warmth from below collected underneath the thick blanket of the super-continent, making it bulge upwards and crack open, like a loaf of fresh bread slowly rising in the oven. Then volcanic hot spots formed in the interior of Pangaea about 200 million years ago. Huge cracks called rift valleys—there's one in north-east Africa—then joined these hot spots together. Over 40 million years, the rift valleys widened, filled with rainwater and turned into lakes, and then oceans. About 160 million years ago, the giant supercontinent of Pangaea split into two parts, one of which was called Gondwanaland. Gondwanaland held all the land that later broke up to become South America, Africa, India, Australia and the Antarctic.

The earth's outer layer seems to be made up of half a dozen or so giant chunky plates. They move between 2 and 10 centimetres per year—roughly the speed your fingernails grow. They slide apart and crash, like maniac dodgem cars without drivers. When two plates collide, one plate dives underneath another plate and heads towards the centre of the earth. This is how our planet makes mountains like the Andes, and the chains of islands, such as the Aleutian Islands that

SUPERCONTINENT CYCLE

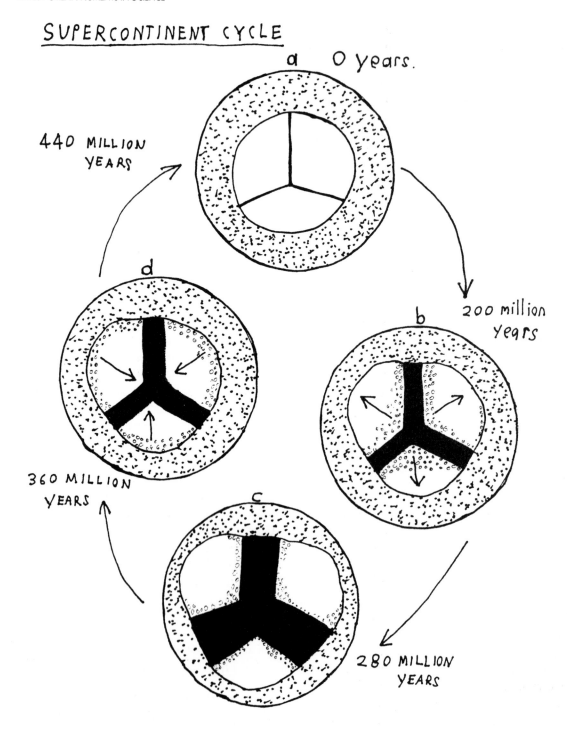

a   0 years.

440 MILLION YEARS

200 million years

b

d

360 MILLION YEARS

c

280 MILLION YEARS

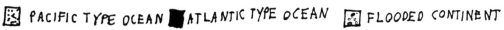

PACIFIC TYPE OCEAN   ATLANTIC TYPE OCEAN   FLOODED CONTINENT

stretch out from Alaska. Right now America is moving away from Europe, and the Atlantic Ocean is getting 2 centimetres wider every year. On the other hand, Hawaii and Japan are rocketing towards each other at a mind-blowing 8.3 centimetres per year.

The plate that carries India is still colliding vigorously with the plate that carries Asia— and as long as this goes on, the Himalayas are still growing, and Mt Everest will get harder to climb. These huge plates are continually wandering over the surface of our planet, carrying the continents and oceans with them. Sometimes the plates grate past each other creating earthquakes, as in the San Andreas Fault in California.

Over the last 200 million years, the fragments of this gigantic supercontinent have now drifted apart almost as far as they can go. Over the next 160 million years, the plates will carry the continents back together again and assemble them into a new supercontinent which should last for about 80 million years. Then the heat welling up from deep inside the earth will start the whole cycle off again.

This brand-new theory about recycling the supercontinents has been put forward by three scientists having separate specialites in the fields of plate movements, oceanography and geochemistry. They are Damien Nance, Tom Worsley and Judith Moody of Ohio, USA. They estimate that there have been six previous supercontinents on our earth, the first about 2600 million years ago, and the most recent 250 million years ago. They also say the Atlantic is just a temporary ocean. It opens up when the current supercontinent splits apart, and then closes when the next supercontinent forms some 500 million years later. Looking at earth through a time-lapse camera, the Atlantic Ocean is a winking eye

that has winked six times in the last half of the earth's existence.

In each cycle, the supercontinent has changed the weather. Last time, the temperature in the centre of Gondwanaland reached a sizzling 50°C during summer, and dropped well below freezing every winter. Gondwanaland was huge, about 40 per cent larger than Europe and Asia combined. As the supercontinents broke apart, vast areas of tropical coastline were created, and the weather became finer and milder.

Now it seems that new life forms appear every time the supercontinents break up. About 2.1 billion years ago, around the time of supercontinental break-up number two, blue-green algae developed the ability to fix nitrogen and survive in an oxygen-rich atmosphere. Without this mutation, there would be no photosynthetic plants today, and there'd be no salad sandwiches.

About one billion years ago, at break-up number four, the very first multicellular animals appeared. Six hundred million years ago, at break-up number five, animals with shells appeared for the first time, and lobsters and prawns began to appear on the menu. And Pangaea, the most recent supercontinent, was famous for the arrival of the reptiles and dinosaurs.

I wonder who will replace us, in 300 million years, the next time around, on board the cycling supercontinent?

REFERENCES

Scientific American April 1987 'The Emergence of Animals' by Mark A. S. McMenamin, pp 84–92
Scientific American July 1988 'The Supercontinent Cycle' by R. Damian Nance, Thomas R. Worsley and Judith D. Moody, pp 44–51
New Scientist No 1672, 9 July 1989, p 16
Scientific American June 1979 'The History of the Atlantic' by John G. Sclater and Christopher Tapscott, pp 120–132

Diamonds

Some are older than our planet and come from outer space. They can improve your car's fuel economy, and if you can afford the space-fare, there's a large one sitting on the surface of Venus, just waiting to be picked up.

They're diamonds and because they're the hardest substance in the known universe, they are used for drilling through kilometres of solid rock, or making super-sharp scalpel blades used in eye operations.

The ancients were fascinated by diamonds. It was the hardest stuff they could find. If you put a diamond on a steel anvil, and then hit it with a hammer, the diamond would embed itself either in the anvil or the face of the hammer. It couldn't be destroyed, unless you were able to hit it along a cleavage plane. Because it was unconquerable the Greeks called it *adamante*, from which we get the word diamond.

Diamond is made of ordinary carbon, but the carbon atoms are arranged in a very special way. They're packed into a very tight crystal structure that looks like a whole bunch of tiny pyramids stacked on top of each other. The only way you can splinter a diamond is to hit it a blow that runs exactly along the faces of these pyramids, along a natural cleavage plane.

If you want to make the toughest grinding wheel possible, you'll have to make it out of diamond. White diamonds are gems and too expensive to use in a grinding wheel, but industrial diamonds, coloured by impurities like nickel and cobalt are much cheaper. To make a diamond grinding wheel, they start off with tiny needle-shaped industrial diamond crystals, smaller than a millimetre. Then they set up a magnetic field, which is arranged like the spokes of a bicycle wheel, running from a central point outwards. The magnetic field acts on the nickel and cobalt impurities in the diamonds and lines up the crystals so they all point outwards. Then they pour a tough resin on the sharp little diamond knives, and you have a diamond grinding wheel that can cut through anything in the known universe.

Scientists have found diamonds inside meteors that have landed on our planet. And since the meteors are older than our planet, the diamonds have to be even older. They are very small, about one-fifth of a micron across—that's about 300 times smaller than the thickness of a human hair. They'd be OK for a mosquito's engagement ring.

But where did these ancient brilliant mini sparklers come from? Those giant hydrogen bombs in the sky—stars. They burn hydrogen and helium to make heavier elements like lithium and carbon, and at the same time give off huge amounts of energy. The astro-physicists think that these tiny diamonds are made in the thin atmospheres of stars as the carbon atoms smash into each other at huge speeds. If this sounds unbelievable, just remember that every atom of gold in the universe was made inside an exploding star.

But you don't need a star to make a diamond. Mother Nature used great heat and pressure to make diamonds 100 kilometres below the earth's surface. Normally we find diamonds inside kimberlites—pipes of solid rock that reach 100 kilometres down. So in the Kimberley mountains in South Africa and Western Australia they've been mining these kimberlites for diamonds.

Chemists have been making diamonds for about 20 years, but it's very slow. It takes a week to make a carat, which is one-fifth of a gram. You probably remember in the Superman movie when he squeezed a lump of coal in his super-fist and turned it into diamond. So the modern alchemists squeeze carbon at a pressure of about 70 000 atmospheres and at temperatures of about 2000°C. In the 1920s, John Logie Baird, the man who 'invented' television, tried to make diamonds. He filled a 200-litre (or 44-gallon) drum with concrete, coal and dynamite. The concrete was there to hold in the force of the explosion, the coal was supposed to turn into diamond, and the dynamite was the power supply. But when he exploded these drums with the dynamite, he never did find any diamonds scattered over the blast area of 100 square metres or so.

But now scientists have made diamonds out of alcohol, or even sewer gas (methane is CH_4). Basically a diamond is just a special arrangement of carbon atoms. And there's lots of carbon in alcohol. So they heat up alcohol at a certain secret temperature and pressure using lasers. This gives a vapour of carbon. Now when steam, which is water vapor, cools down, it makes water. And when

carbon vapour cools down, 99 per cent of it turns into graphite, but 1 per cent turns into diamond. The deposited diamond has a slightly different structure, and it is slightly harder than natural diamond—this is because the new structure is an amorphous (noncrystalline) structure. They can make this 'diamond dew' deposit on any surface they like, and it can be laid down at 0.3 to 3 microns per hour.

So if they put a thin coating of diamond on a motor car engine bearing, the bearing will never wear out because diamond is the hardest stuff in the known universe. And if they put a thin layer of diamond on the periscope of a submarine, the front lens won't get abraded by the sea water. Same for plactic sunglasses.

Because diamond is such a good conductor of heat and electricity, they want to grow layers of it in semi-conductors. At the moment, semi-conductors will operate at fairly low temperatures. But if they grow layers of diamond for semi-conductors, they could run not just at 100°C , but at 5000°C.

The electronics industry is already making experimental chips from diamond. Diamond is a semi-conductor, and these diamond chips have several advantages over silicon chips. They can keep operating at higher temperatures and speeds than silicon chips. Sony has already released tweeter speakers that make the sound with a very thin layer of diamond.

But diamonds can also improve your fuel economy. The first few thousand kilometres in a new car is probably the time you should do your most careful driving. Think of running-in a car engine like polishing silver—medium to high speed on the engine but not too much pressure on the accelerator. Engineers have found that adding a carat of crushed diamonds to the engine oil will help the engine wear in properly. So you can get 20 per cent extra fuel economy for the life of

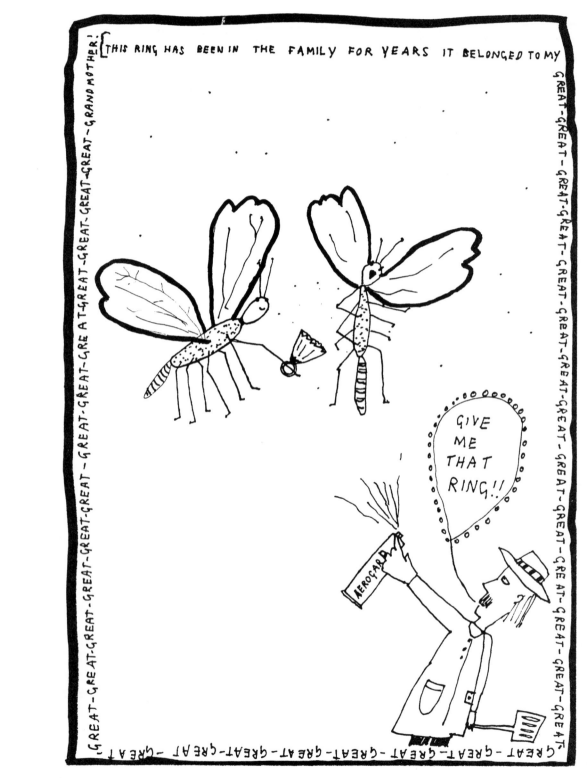

that engine. But it will cost you—a carat of diamonds sells for from $6 for an industrial, up to $6000 for a gem. So diamonds are an oil's best friend.

But if you want to get a free block of diamond, there's a large superb gem, the size of a 10-cent piece, sitting on the surface of Venus. When the Soviets dropped their robot landing craft onto the surface of Venus, they wanted to be able to take photographs for the half hour or so that the spacecraft would survive after it landed. The conditions are terrible—a temperature of 500°C, an atmospheric pressure 100 times our own and an atmosphere of carbon dioxide and sulphuric acid. That atmosphere is enough to chew through the spacecraft, as well as the front lens of any camera. So the Soviets used a large block of top-grade gem-quality diamond as a window to protect the lens. So if you want a free diamond, you know where to get one—from Russia with love.

REFERENCES

New Scientist 18 February 1982, p 440
Scientific American May 1985, pp 18–19
New Scientist No 1520, 7 August 1986, p 25
Nature Vol 332, 23 April 1988 'Formation of Xe-HL-enriched diamond grains in stellar environments' by Uffe G. Jorgensen, pp 702–705
Popular Science September 1988, pp 58–60, 90
Planetary Report Vol IX, No 1, January/February 1989, pp 9–13
Discover July 1989, p 12
New Scientist No 1603, 10 March 1988, pp 50–53

Dirty Air

*M*ore than 80 per cent of the world's city dwellers breathe air that's too dirty—that's about one-quarter of the world's population.

And the dirty air is just not a nuisance or an inconvenience, it kills millions of people and costs us billions of dollars. Mexico City is the world's largest city with a population of 18 million. Their dirty air is so filthy that it's like smoking two packets of cigarettes per day. The dirty air has killed 600 000 people over the last six years. In Mexico City, one citizen is air-polluted to death every five minutes.

Our planet has always had natural chemicals of pollution, but has also had balancing mechanisms to deal with these chemicals. Volcanic eruptions, decaying vegetable matter and sea spray release more sulphur than all of the power plants and heavy industries in the world. Bolts of lightning create oxides of nitrogen, just like industrial furnaces and automobiles. And the famous blue in the Blue Mountains near Sydney, comes from chemicals called terpenes oozing out of the eucalyptus trees.

For hundreds of millions of years, these chemicals have been travelling through the ecosystems of our planet. Chemicals come from plants and animals, return to the earth, float to the ocean, and are lifted into the clouds by evaporation before they enter the cycle again. Your average atom of oxygen takes about 2000 years to go through the cycle once. You could be breathing some oxygen atoms that were last used by Jesus Christ.

But the human race expanded enormously, and changed this natural balance. As soon as you get enough thoughtless humans in one spot, you get pollution.

In the 1600s in London, people used to throw household garbage, as well as chamber pots of urine and poop, into the streets every night. And every night, the pigs would be let loose to scavenge the streets clean again. This was at a time when the Chinese and Japanese had garbage collections, public baths and paper money.

New York city in the 1890s was roughly the size of present day Melbourne, and they had a massive pollution problem. Each day, the horses of New York city dumped hundreds of tonnes of solids on the streets, as well as a quarter of a million litres of urine. That's enough urine to fill 100 backyard swimming pools—you can imagine what that would smell like on a hot day. And if you've had trouble getting roadside service on your car, imagine how you would get rid of a dead horse.

But it was the twentieth century before we humans really succeeded in polluting the very air that we breathe on a world-wide scale. Acid rain is destroying forests across the world. We've dumped enough carbon dioxide into the atmosphere to start off the green-house effect—just think what that will do to

real estate around Surfers Paradise. It took just 35 companies only 50 years to make enough chlorofluorocarbons to punch a hole in the ozone layer—and at the very least, that will mean millions of extra cases of skin cancer. We are dumping 65 000 new chemicals into our environment each year—and we don't even know what they do.

Take just one polluting chemical, such as lead. Most of us take in lead everyday, from water plumbing pipes, car exhausts and old paint. It's a new problem. Peruvian skeletons 2500 years old have less than a thousandth the amount of lead in their bones that we have today. Each car running on leaded fuel, a 'lead sled', spews about a kilogram of lead from the exhaust pipe each year. About 85 per cent of this lead gathers in the bones, brain and kidneys. Twenty per cent of all American preschool children have excessive levels of lead in their blood—and the percentage is higher if they live in the inner city, or near an expressway. Too much lead lowers children's intelligence and shortens their attention span, making it hard for them to learn new things.

Too much lead can give adults high blood pressure and this can cause heart attacks. In America, lead-free petrol was introduced in 1976. Over the next four years, the amount of lead in the blood dropped by about one-third. The Environmental Protection Agency calculates that this drop will prevent 65 000 heart attacks each decade. Maybe the human race will recover some of the intelligence that it has lost.

But there is a bright spot on the horizon. Los Angeles is going to clear its dirty air by the year 2009. Smog, that mixture of smoke and fog, first appeared above Los Angeles around World War I. By World War II, the smog would sting your eyes. And by the mid-1950s it damaged the food crops and made breathing difficult. In 1989 a group of air-quality officials studied the problem. The group included an economist who gave a dollar value to each smog-related disease. They found that if they reduced the smog, they could save $12 billion in health-care costs.

In the first five years of the 20 year plan, they are banning aerosol hairsprays and deodorants, and making companies install the most efficient anti-smog equipment available. The next stage will be to try to convert motor vehicles to burn cleaner fuel, such as methanol, and encourage people to use car pools. And by the year 2009, any vehicle that travels on the roads of Los Angeles will have to be powered by electricity. But it will be cheap—it will cost $3 billion dollars a year to save that $12 billion in health-care costs.

The real problem has been that we have always thought of the environment as being free—so it's been OK to dump our waste in the gutter of the street, on the nature strip outside our house or into the nearest creek. But this year, for the first time, Sweden is assigning a cash value to its environment. Fresh air and clean water are now treated as an asset in the economy. If a lake is polluted by a chemical company, then so many krona are subtracted from Sweden's gross national product. Perhaps in the next ten years, the rest of the world will realise that environmental destruction costs an awful lot.

REFERENCES

National Geographic April 1987 'Air; An Atmosphere of Uncertainty' by Noel Grove, pp 502–537
Car Australia April 1988, p 123
Time 27 March 1989, p 51
Scientific American December 1989, p 11

Dolphins and Whales Stun their Prey

*i*n 1983, a sperm whale accidentally strayed into a New York marina.

Two divers went down to have a close look at it, because a sound was coming out of a circular spot above the whale's upper jaw. The sound was so intense, that their hands were forced away from that spot on the whale, like a water cannon pushes away a political protester!

We know that sperm whales, killer whales and dolphins all use sound to find their food. They have a sonar system: they emit sound waves that bounce off their dinner, and then they measure how long it takes for the echo to come back. The longer the time for the echo to come back, the bigger the distance. The noise starts off in their blowholes. Dolphins and whales both have blowholes on the tops of their heads. It's their equivalent of a nose. Dophins also have a large blob of fat, called the 'melon', just above the upper jaw. The melon focuses the sound from the blowhole, intensifies it, and aims it in a particular direction.

When whales and dolphins run their sound generator at low volume they can use it to find their dinner. But it seems that they can crank up the output to 'warp factor 9' and emit powerful blasts of sound that can stun fish, and even kill them. This sound lasts for about 200 microseconds and sounds like gunfire or machinegunfire. It's quite different from the normal clicks and squeals that dolphins use.

It's a little hard to measure just how much sound these mammals can put out in captivity. Making the full-volume killer noise in a swimming pool would be like letting off a hand grenade in a car while you were still inside. But in the ocean, dolphins have been measured putting out 230 decibels of sound underwater. That's a lot of noise, when you think that a hard-core thrash band puts out 120 decibels, while a jumbo 747, running flat out, can generate only 140 decibels. So 230 decibels is like tossing dynamite into a goldfish bowl—it's as loud as one billion jumbos, running flat out, all together in the one spot.

This work was done by Kenneth Norris at the Long Marine Laboratory at the University of California at Santa Cruz. He calls it the 'Big Bang' theory. And it answers a lot of questions.

In 1980, Norris watched three spinner

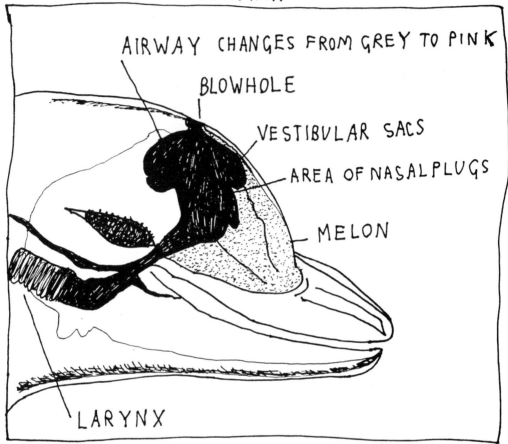

DOLPHIN

AIRWAY CHANGES FROM GREY TO PINK

BLOWHOLE

VESTIBULAR SACS

AREA OF NASAL PLUGS

MELON

LARYNX

dolphins in a large marine park tank swim past a school of 20-centimetre long fish 24 times, each time hitting the fish with the low-level sound. The fish were so disoriented, that they could not keep themselves upright or stay in the school formation, and one of them even got sucked down a drain.

Fishermen off the coast of Argentina and Mexico have seen anchovies being herded together by dolphins. The anchovies became so tired and lethargic that the fishermen could actually pick them up out of the water with their hands. And while they were doing this, the dolphins were making runs through the anchovies, grabbing a mouthful each time.

So this would explain how sperm whales can catch squid, even though the squid are much faster than the whales. And it would also explain how sometimes squid swim out of the mouth of a freshly killed whale, and show no signs of tooth marks.

And maybe that's why dolphins are so polite to each other and switch off their sonar gear when they get close—they are all carrying loaded guns.

REFERENCES

New Scientist No 1388, 15 December 1983, p 807
Scientific American October 1987, p 23
New Scientist No 1532, 30 October 1986, p 40–43
New Scientist No 1586, 12 November 1987, p 32
Discover February 1988, pp 81–83

Doughnut Submarines

*i*n the twenty-first century, there'll be a new generation of submarines. They'll be nuclear-free, but they will still be able to go around the world without coming up for a breath of air. And they'll be made of very special doughnuts.

Every navy would love to have one of those nice, nuclear-powered submarines that carries Trident Missiles and can launch 288 nuclear warheads against 288 separate targets within just a few minutes. That is enough to wipe out every single city in the Soviet Union with a population of more than 100 000. But these submarines are just too big and expensive. They weigh 19 000 tonnes and cost over $1 billion. The beauty of a nuclear-powered submarine, and there aren't many beautiful things about them, is that they can run underwater indefinitely at speeds of 50 to 70 kilometres per hour.

The World War II-type diesel-electric submarines—and everybody's still using them—are a lot slower and easier to find. The main problem with diesel engines is that they burn up a lot of air. If the hatches were shut, the crew and the engines would suffocate within minutes. The engines charge up huge banks of batteries that run an electric motor to turn the propeller. However, these batteries are flattened by two or three days' running at low speed, or an hour's running at full speed. In the old days, every time the engines were started up, the sub had to come up to breathe, and risk attack. But during World War II, the Germans invented the snorkel, so the submarine could breathe without surfacing. But even a snorkel isn't good enough these days—satellites can pick up the tiny trail they leave in the water.

The Italians have come up with a new design for the old-fashioned diesel-powered submarine, so it doesn't need to come up for air. It is the brainchild of Giuno Santi, the vice-president of an Italian defence company. Like many good ideas, it is brilliantly simple.

The conventional submarine is in the shape of a long cigar. Its hull has thousands of steel plates that are joined onto a frame. Santi's idea was to make the submarine out of hollow steel doughnuts stacked side by side. He started off with a steel tube with the diameter of a soccer ball and walls as thick as your

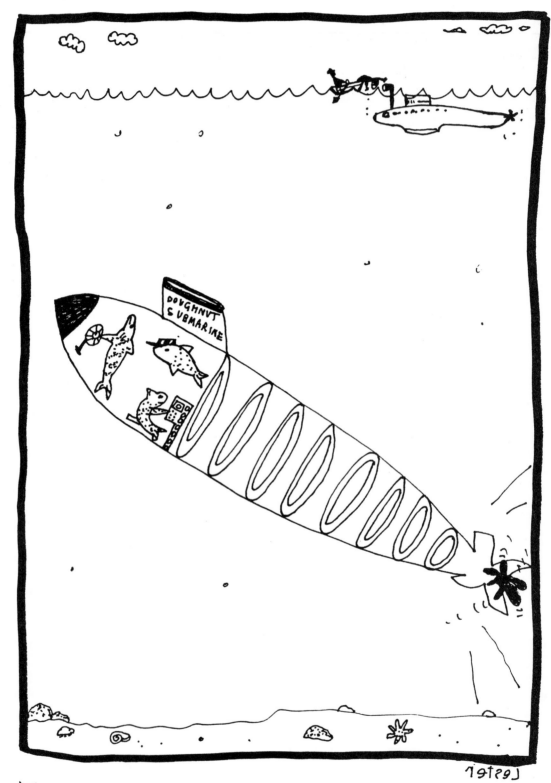

finger. He then bent the tube into a circle, and welded the two ends together to make a doughnut.

The front end of the submarine starts off with a small doughnut about 0.5 metre across. Then another doughnut a little bit larger is welded onto that, and then another slightly larger doughnut welded onto that again. The doughnuts get bigger until they get to the belly of the submarine. Then the doughnuts get smaller towards the back end. So the sub looks like the Michelin tyre man laying down.

It's an unusual way to build a submarine, but there are bonuses. First, you can store pure oxygen inside each of the doughnuts at a high pressure (about 350 atmospheres). This oxygen is for the engines and crew. Secondly, the doughnut hull is five times stronger than one made out of bent steel plates. Not only that, it is also much easier to get the perfect hydrodynamic teardrop-shaped sub with steel tubes. All this gives you a self-sufficient efficient submarine.

The first submarine made out of these doughnuts is about 9.5 metres long—about as long as a flatbed truck—looks just like a huge steel pear on its side and displaces 29 tonnes. It carries a crew of four, has an underwater range of about 200 nautical miles at 6 knots and can dive to 400 metres. That's like jogging across Bass Strait without coming up for air. It has 30 times the range of a old-fashioned midget sub, and can dive four times deeper.

And when you make a bigger sub, it can carry more oxygen in the extra doughnuts, and that means that it can go further. The Italians are now working on a sub 48 metres long. This will travel 8000 nautical miles submerged at 5 knots—that's once across the Pacific Ocean. If they made a sub 70 metres long, you could store enough oxygen inside the doughnuts to travel around the world, completely submerged, at a reasonable jogging speed of 8 knots.

The whole point of a sub is secrecy: the motto of the Royal British Navy is *Venio Non Videor* which means 'I come unseen'. If the sub has to come to the surface, you may as well have a ship. Nuclear-powered submarines, first introduced in 1955, were the first invisible submarines, but they are very expensive, and these days, not every harbour will permit them to enter. But soon any country can have nuclear-free invisible doughnut submarines. And with their doughnuts, they'll be able to run rings around the rest of them.

REFERENCES

Military Technology November 1988, pp 61–70
New Scientist No 1658, 1 April 1989, pp 34–37
New Scientist No 1662, 29 April 1989, p 55

Electronic Smog

*i*t's not hard to find chemical pollution, noise pollution, heat pollution and even visual pollution. But now there's a new type of pollution that has already penetrated to every corner of the globe.

You can't smell it, you can't see it, you can't feel it or hear it. It's coming out of billions of electric and electronic devices from all over the planet. It's called 'electronic smog'.

Electronic smog is wild electromagnetic radiation, usually radio waves that have escaped from their enclosure to drive other electronic devices and their users crazy. Like a radio signal, you pick up electronic smog with an aerial—but the aerial can be part of the metal case or a small length of wire inside, or even the 240-volt power cable that runs the device.

The most common form of electronic smog is television interference. It scrambles reception on your favourite shows when the neighbours start using power tools, cars with faulty ignition systems drive past, or the CB radio down the road goes on the air.

But in Germany, electronic smog has shut down an autobahn. The US military set up a big radar unit near the old Berlin Wall to spy on the Soviets. But the radar unit was also spraying a nearby autobahn with microwaves. The radar energy was entering cars on the autobahn either by going straight through the glass in the car windows, or by being picked up by the metal panels of the body. Unwanted currents and voltages were

set up in the wires that ran through the cars. Those floating currents tricked the delicate micro-electronic circuits into switching on and off.

In the good old days when cars were simple and electronics hadn't been invented, the only electrical appliances on your car were the lights, the wipers and maybe a valve radio. These appliances all ran on very large currents. So a tiny current from your local military radar unit or radio station didn't have much effect.

But now the cars have fancy electronic gadgets that are all controlled by tiny voltages and currents—antilock braking, electronic automatic transmission, fuel injection, electronic ignition, AM/FM stereo radio cassette, CD disc player, cruise control, burglar alarm, electronic door locks, telephone, and even four-wheel steering and computer-controlled active suspension.

Whenever the radar unit was switched on, a traffic jam full of dead cars suddenly appeared on the the autobahn, as electronic ignitions stopped igniting, and fuel injection units stopped injecting. Even the car radios had sudden memory blackouts, and forgot all of their driver-programmed stations. The military had to build a giant wire cage, called a Faraday cage, over the autobahn to keep the traffic moving.

But the electronic smog can also be used to spy, and invade your privacy. Anyone with the right equipment can pick up the signals from your personal computer from a range of several hundred metres. It's easy to prove for yourself that your computer is on the air—hold a portable radio near the screen and listen.

When your computer handles data, hundreds and thousands of electrical signals flow rapidly inside the computer. Some of these signals are amplified to tens of thousands of volts, so you can see them on the screen. They are fed through wires that act as radio antennas, and broadcast everything appearing on the screen. You can pick it up half a kilometre down the road.

Electronic smog nearly stopped the construction of the new King Fahd International Stadium in Saudi Arabia, when the stadium started burning the workers who were building it. This stadium has 24 hollow steel masts, which are each 60 metres tall. They are arranged in a circle 290 metres across and carry some 20 kilometres of steel wire. That enormous length of steel wire acted like a really good radio aerial. By a terrible coincidence, only 3 kilometres away was a radio station broadcasting at 1.2 megawatts—that's about 60 times more powerful than a typical radio station. Because the stadium was so big, and so close to such a powerful transmitter, a huge voltage was induced in the wires—4 volts per metre of wire. Multiply that by 20 kilometres of wire and you get high tension lines. Fluorescent lights that were not even connected to the power switched themselves on, and flickered in rhythm to the radio signal. Electric sparks jumped from girder to girder, and actually sang in rhythm with the radio signal. Riggers working on the building were burnt by the electric sparks, and had to wear safety harnesses because of the danger that a small electric shock could throw them off balance to their death.

The problem was solved by electrically separating the entire structure into short sections, and then earthing each section independently. Even the plumbing was broken into sections with lengths of rubber hose. All the cables carrying audio signals were shielded by wrapping them in aluminium tape. And cables carrying very low levels of signals such as video signals from broadcast and surveillance cameras had to be replaced by optic fibres, which run on light.

Electronic smog has thrown aeroplanes out of the sky. In 1984, a Tornado fighter of the West German Airforce flew close to the high powered transmitters of Voice of America and Radio Free Europe near Munich. This aircraft is totally controlled by computers operating at low voltages. It crashed, killing both crew members. In the United States, several army Blackhawk helicopters crashed when they flew too close to powerful radio transmitter antennas. So now the US Airforce has a list of 300 powerful transmitters that pilots must avoid, but for some unknown reason, the list is classified.

In fact, the United States Airforce Space Command is one of the worst offenders. They built a very powerful $90 million radar station only 2 kilometres from the approach end of the Robins Airforce Base runway in Georgia. This radar is an early warning unit, so it looks for the submarine-launched missiles of a possible Soviet sneak attack.

But US Airforce engineers were worried that this radar unit could trigger electrically controlled explosive devices on the military aircraft. Explosive devices are used right through military aircraft—to work the ejection seats, to dump fuel from the tanks, or to fire the rockets. So whenever a plane

comes in to land, the $90 million early warning radar unit is shut down, and America's defence against a Soviet sneak attack is compromised for 90 seconds.

Electronic smog can be so powerful that it can kill. In fact, the downward-looking radar units of some jet fighters are so powerful that if the pilots switched them on while flying close to the ground, they would kill rabbits, and severely disorient humans.

The Raytheon Corporation of the USA is working on a jet fighter that will jam enemy communications by blasting out a super-powerful broadband radio signal—4 megawatts of power. In fact, the jet fighter has to carry four extra engines just to generate enough power to run this jamming unit. Four megawatts would have been enough to cook the pilot like a chook in a microwave. Your average microwave oven runs at only 500 watts—this is 8000 times more powerful. So to protect the pilot, they had to coat the transparent canopy over the pilot with a very thin layer of gold, to stop the electronic smog getting through.

Pilkington Brothers, who make glass, now sell glass that is coated with an incredibly thin molecular layer of indium tin oxide. The layer is so thin that it will let light through, but it blocks radio signals between 30 megahertz and 10 gigahertz. You can use this glass to protect the privacy of your computer, and to stop general electronic smog getting either in or out of your windows. But if you want to protect the rest of your house, you should build it from metal, or if the budget's tight, cover it with alfoil.

This pollution can only get worse. Plastic does not stop it—and most electronic devices have plastic cases, and increasing amounts of plastic are being used in cars. The European Common Market is setting up automobile standards for 1992 that will recommend that all wiring for electronic control systems runs inside shielded cable, and that all car computers sit inside earthed metal boxes. Where possible, old-fashioned copper wires will be replaced by optic fibres, which are smog free. They don't pick up incoming electronic smog, and they don't produce it either.

At least we don't think so, but when we use optic fibres for underwater telephone links, sharks come from kilometres around just to have a feeding frenzy and munch through the optic fibre until it stops working. It looks like the sharks have their own way of dealing with electronic smog down on the ocean floor.

REFERENCES

New Scientist No 1534, 13 November 1986, p 28
New Scientist No 1549, 26 February 1987, p 40
New Scientist No 1607, 7 April 1988 'Electronic Smog Fouls the Ether' by Barry Fox, pp 34–38
Time 7 February 1989, p 31
Car Australia August 1989, p 111

The Eternity Aeroplane

*t*he Canadians have built the world's first plane that can fly forever.

The weird thing is that it doesn't carry any fuel. It's like a kite, but there are no strings attached. It doesn't use the wind, but gets its power from an energy beam that is blasted up from the ground. And it will be soon used as a cheap alternative to communication satellites.

There is a long history of people who tried to transmit electrical power through radio waves. Heinrich Hertz, the first man to generate and pick up radio waves, the very man after whom megahertz is named, tried to broadcast electrical power through radio waves, and failed. And Nikola Tesla, who actually invented the radio before Marconi, the standard AC electrical motor and the distribution system that made electricity available to all, tried to transmit electrical power through radio waves, and he failed too.

They both failed because they tried to work with ordinary radio waves which are too long. An average AM station has waves that are about 400 metres long, while the waves from a good FM station like 2 JJJ are about 2.8 metres long. It's hard to focus such big waves into a tight, sharply focused beam.

If you want to aim a radio beam accurately, you need microwaves. Microwaves are just radio waves that are short. They have a wavelength of about 10 centimetres, instead of a few metres. Microwaves can be focused very accurately because they have such a short wavelength. In fact they are used to transmit TV signals and telephone conversations around Australia. The beams have to be aimed very accurately to hit a tiny dish 30 kilometres away on the horizon.

Microwaves were invented in World War II, with radar. Radar uses microwaves, and so does your microwave oven at home.

But the first steps towards microwave-powered flight happened in 1964. William C. Brown of the Raytheon Corporation tied a specially modified small helicopter to the ground with a rope 20 metres long. On the outside of the helicopter were special antennas, an array of 3000 point-contact semi conductor diodes. They picked up microwaves from a transmitter dish on the ground, and turned them into DC electricity. The DC electricity powered a 270-kilowatt motor, which turned the helicopter blades.

As long as he sent microwaves up to the chopper, he could keep it flying. But the special antennas were too heavy and so the microwave-powered flying machine stayed in the laboratory as an impractical experiment for more than 20 years.

In 1982, a new type of antenna was invented. It was 10 times lighter. All of the circuitry was etched into a thin plastic film, just like in printed circuit boards. In 1987, William C. Brown used this new antenna to achieve his dream. This time around, he built a plane with a 4.5-metre wing span. And it was very light—only 4.5 kilograms. There were little nickel–cadmium rechargeable batteries inside the plane. They drove a very efficient samarium–cobalt magnet motor, which then drove the 60-centimetre propeller. The propeller was a very efficient slow-speed model based on the propellers used on human-powered aircraft, like the one that flew across the English Channel.

On the day of the big test flight, the nickel–cadmium batteries in the plane had already been charged up. William C. Brown held the plane in one hand and switched on the batteries. The propeller began to spin, and once it had reached full speed, he pushed the plane into the air and it slowly climbed away from the ground. It was controlled with a standard amateur aircraft remote-control kit. At this stage, it ran on the internal rechargeable batteries.

The test flight was carried out next to a 5-metre parabolic dish that looked like a giant wok. It was the ground transmitter, and it put out 10 000 watts of electrical microwave power—as powerful as 20 microwave ovens.

At a height of 100 metres, the plane ran into the microwave beam coming from the ground. At this stage, the batteries switched off, and the plane was powered only by microwaves. But the 10 000 watts had to travel through the air, be picked up by the special antennas, and then fed to the special samarium–cobalt motor.

There were big energy losses. Air isn't the most efficient way to send electrical power.

So by the time this 10 000 watts got to the electric motor, there were only 150 watts left—only 1.5 per cent of the original transmitted power. But this was enough to lift the plane up to 150 metres where it flew lazily in a circle in and out of the microwave beam for a few hours. It cruised happily at about 35 kilometres per hour, as the microwaves recharged the batteries.

But this plane was not big enough to carry a communication dish—it was just a model. The plane that carries the communication dish will have a wing span of over 40 metres. It will be powered by a circular transmitter roughly as big as a football field. The ground transmitter will put out 500 000 watts of power, but only 30 000 watts will get to the motor. But this will be enough to fly the plane in large 4-kilometre diameter circles. It will travel at a speed of 180 kilometres per hour. This is the speed that you need to fight the jet-stream at an altitude of 20 kilometres.

The plane itself will cost about $2 million while the ground station will cost about $50 million. But this is about one-third the cost of launching a single communication satellite. That would cost about $100 million for the satellite, and another $40 million to get it into orbit.

The plane can be used for direct-broadcast television, like Sky Channel, and car phones. A plane flying at an altitude of 20 kilometres could broadcast to an area about 600 kilometres in diameter. That is the distance between Sydney and Albury. So you could have direct television broadcast to rural areas with a home dish the size of a garbage can lid.

The plane can carry a military radar to look for invaders. It could survey the ocean to look for people fishing inside our 200-nautical-mile fishing limit, and it could look for drug smugglers and illegal immigrants. It could also be used for research—to monitor the level of carbon dioxide, aerosols, and trace

gases in the atmosphere. It could look at clouds and the amount of energy coming from the sun.

People flying inside metal aeroplanes would be protected from the cooking effects of the microwaves by the metal skin of the plane. But migrating birds have no protec-tion, and wild geese could soon be falling, fully cooked, out of the skies.

REFERENCES

IEEE Spectrum September 1987, p 19
New Scientist No 1585, 5 November 1987, p 35
Popular Mechanics December 1987, p 55
Popular Science January 1988, pp 61–65, 106, 107

Falling Cats

*n*ow we all know that cats have nine lives. And a recent experiment proves it. Cats can fall from a 32-storey building and survive!

The scientific experiment to prove this was done in New York where 95 per cent of the people live in high-rise buildings. So the combination of people, cats, high-rise buildings and concrete pavements gives you a ready-made experiment that can test the aerodynamics and impact resistance of cats.

It's important to point out that the cats were not deliberately thrown out of high-rise buildings. They jumped out by themselves. At least, that's what the owners of the cats said when they took the cats to the hospital. The hospital was the New York City Animal Medical Centre. Over a five-month period, two vets at the hospital, Wayne Whitney and Sheryl Mehlhoff, looked at fallen cats—132 of them. One hundred and thirty two cats fell from a height of a least two storeys. So it was raining cats at about one cat per day.

These vets analysed the figures. The average fall was five and a half storeys. Ninety per cent of the cats survived. One out of every four cats falling from two storeys broke one bone in its body. And when they looked at the cats falling from seven storeys, there was an average of one broken bone per cat. So for roughly four times the height (seven stories versus two storeys), there were roughly four times as many broken bones. All very reasonable.

But then the statistics showed something very funny. As the cats fell from greater and greater heights, they broke fewer bones. One in every 10 cats falling from 32 storeys broke one bone in its body. So it was safer for a cat to fall from 32 storeys than it was from seven storeys—in fact, two and a half times safer!

Now gravity sucks you down towards the ground. You fall down a gravity well. If there was no atmosphere, you would just keep on falling faster and faster. But we do have an atmosphere. You can feel wind resistance when you foolishly put your hand out of a car window. Wind resistance slows you down. There is a balance between the suck of gravity and wind resistance. When the wind resistance balances out the suck of gravity, you are now travelling at what is chillingly called 'terminal velocity' and you don't fall any faster.

This balance is different for each animal. For a heavy human being, the terminal velocity is about 200 kilometres per hour. But a cat is lighter and fluffier, and the terminal velocity for a cat is only 100 kilometres per hour.

Now we humans are hopeless when we fall out of buildings. Hardly anybody can survive a fall of more than six storeys onto hard concrete. And the usual causes of death are

head injuries, and bleeding from inside the gut. A terminal velocity of 200 kilometres per hour is bad news.

Cats have a lower terminal velocity, but that is only part of the story. Cats are magnificent athletes, and when they fall, they immediately twist and turn in space so that all four feet are pointing downwards. Adult humans on the other hand are hopeless and just tend to tumble, while babies, with their big heads, tend to land on the top of their skull.

Now it turns out that a cat reaches its terminal velocity after about seven storeys. So here is the theory as to why there are fewer injuries when cats fall from great heights. While the cat is still accelerating it is a little bit tense. If it happens to hit the ground while it is tense, it will be more likely to suffer injuries. But once it reaches its terminal velocity it goes into a state of dynamic tension, a bit like Charles Atlas. It has all of its legs stretched out quite horizontally, and it looks a bit like Rocky, the famous flying squirrel.

This means that it is more likely to spread the impact over its body, not just on its legs, but on its chest as well. And if it has a state of dynamic tension in its legs, it can spread the impact of the fall in much the same way that a parachutist does, by landing with the legs and hips bent, and then rolling forward.

In fact cats are so in tune with their bodies, that the cat that fell 32 storeys onto the concrete pavement had only a slightly collapsed lung and a chipped tooth and was sent home after a few hours.

But maybe the cats didn't jump after all— maybe they were pussed. So if you fall out of a tall building, make sure you take a cat with you to show you what to do.

REFERENCES

JAVMA Vol 191, 1987, pp 1399–1403
Nature Vol 332, 14 April 1988, pp 586–587
Discover August 1988, p 10
Science Digest August 1988, p 22

Falling Fish and Flying Saucer Nests

*t*here have never been any really reliable reports of it raining cats and dogs, but there have been lots of frogs, fishes and various shellfish coming down by the bucketful.

And they got there by riding the whirlwind—the same whirlwind that brings us mysteries from outer space.

There are dozens of reports of falling sea creatures. In Yorkshire in 1844, people held out hats during a rainstorm to catch the falling frogs. In 1984, residents of Canning Town, UK, found some 35 fish scattered over their gardens one night after a rainstorm. The Natural History Museum in London identified the fish as the sort found in the Thames River, just a few kilometres away.

But a few kilometres is nothing for these fishy falls. In mid-1984, a garage owner in North Yorkshire near Thirsk found starfish and sea snails covering his garage and driveways. The sea snails were still alive. They must have had a rapid trip to his garage—Thirsk is 45 kilometres from the ocean.

But creatures from the ocean have been carried for even greater distances. In 1983 the inhabitants of the small French village of Dilhome found thousands of tiny sea shells scattered across an area roughly the size of a swimming pool. They actually saw them land during a thunderstorm. Yet the village is about 80 kilometres away from the ocean.

But it is not just small objects that can get carried from the ocean. In Brighton in the United Kingdom in 1983, Julian Gowan was outside his house when a large crab fell from the heavens. It was dead on arrival, and weighed about 125 grams—that's roughly the weight of an iceblock. A few minutes after the spider crab dropped from the heavens, severe thunderstorms swept across his house.

In February 1989, the residents of the Queensland town of Ipswich received a late Christmas present of thousands of sprinkling sardines from God. And in January 1990, inhabitants of Jerilderie in NSW, were invaded by fish raining down from heaven.

DINNER!

The cause of all these seafood showers is the mini whirlwind, like the one that carried Dorothy to the Land of Oz, in the movie *The Wizard of Oz*. You often get a whirlwind, or vortex of swirling winds, rushing around immediately underneath a rapidly growing storm cloud. Usually the vortex is invisible, but if you look carefully, you can sometimes see a spout looking like an ice-cream cone reaching down from the bottom of the cloud to the ground.

If the cloud is over land, it can form the mighty and terrible tornado, often seen in the southern and midwestern areas of the USA. In a tornado, the vortex is usually several hundred metres across, with winds travelling between 100 and 300 knots. We're lucky we don't get the many of the big industrial-grade winds here in Australia—we get the much milder willy-willy.

But if the vortex of wind is not over land but over water, it forms a waterspout. Waterspouts can be tricky—often you can't see them stretching down from the clouds. Sometimes, a circular patch of white water is the only warning the crew of a yacht gets before they enter the violent winds of the whirlwind. And there's no warning for a school of fish close to the surface—they can be picked up by the waterspout and carried hundreds of kilometres away before being dumped.

But these rotating winds, these whirlwinds, don't just rain seafood in biblical proportions—they enrich our crank culture by making so-called 'flying saucer nests'. You've probably read about them on the front page of your local afternoon newspaper. These flying saucer nests are flattened circles in the fields, usually in long grass or wheat, oats or corn. They're between 3 and 30 metres in diameter, and sometimes there are signs of burning. The circles can occur by themselves, or in groups. Usually they're on the lee side of an isolated hill, away from the prevailing wind. They can be as far as 6 kilometres downwind of the nearest hill. There is often a very sharp edge between the flattened inner part of the circle, and the untouched area outside the circle. The grass or crops can be pushed flat in either a clockwise or anti-clockwise direction.

In fact, these circles in the corn are not an overnight sensation—they go back to the Middle Ages. The superstition of that time blamed the circles on 'mowing devils'. The story goes that a farmer asked a labourer for a quote to mow his field. The labourer came back with a quote that the farmer thought was too high. The farmer was so angry that he said that the devil would mow his field before he would give the job to the labourer. The very next morning, the farmer awoke to find that his field was covered with circles, as though the devil had taken up the farmer's offer.

But today, people have disregarded this story. They are convinced that passing flying saucers, wanting to land on our planet, but being shy of the TV cameras, would land in a field, and leave behind obvious clues. Terence Meaden of the Tornado and Storm Research Organisation in the United Kingdom, became intrigued by the so-called flying saucer nests. So over the last decade he examined hundreds of them and even found witnesses who had seen them appear right in front of their very eyes.

This is unfortunate for the people who believed in the quite exciting theory that a couple of little green persons had borrowed their parents' flying saucer to fly to the nearest backward planet to terrorise the residents, and had left a few flattened circles where they had landed.

One witness saw a flattened circle appear while he was out walking on a summer evening. First he saw the long stalks of corn bend over and then straighten up again. It was as though an invisible force was rolling over the corn at about 80 kilometres per hour. Suddenly, the force stopped in one spot and in just 4 seconds swept out a circle about 25 metres across. There was a loud hiss and a rustling noise as the corn was flattened out.

Meaden reckons that a mini whirlwind or vortex caused the flattened circle. These vortices are invisible in air, but not in water. It's quite easy to see a vortex develop in your nearest creek or river. It usually happens along the boundary between two currents that are either moving at different speeds or are at different temperatures. A vortex can also happen when a current splits into two parts to go around a rock or a tree trunk, and then joins up again downstream. The vortex appears where they join up.

Meaden now has a theory to explain what the eyewitness saw. It's based on experiments with smoke rings and computer simulations. As the wind came over the hill, a small vortex of spinning air formed reaching up a few hundred metres above the ground. Spinning columns of air are unstable, and soon decay. But before they decay, they go through a stage where a bulge forms half way up the column. Inside this bulge is a doughnut-shaped mass of spinning air, which moves down to the ground as the column of air decays. When it finally does hit the ground, it expands outwards and creates the famous flying saucer nests. When a descending doughnut hits the ground hard, the soil under the crops can be compressed. But sometimes, the spinning mass of air brushes the crop gently, and doesn't fully flatten it.

Meaden also reckons that friction between the moving air and the surrounding air generates a plasma of hot gas. This would account for the hissing noise, the sharp edge between the inside and outside of the circle, and the occasional burn marks.

Now these vortices might not bring flying saucers to our planet, but they do home-deliver seafood cocktails.

REFERENCES

New Scientist No 1615, 2 June 1988, pp 38–40
Daily Mirror 7 February 1989, p 1
Journal of Meteorology Vol 14, No 140, July/August 1989 'Circle Formation in a Wiltshire Cereal-Crop—An Eyewitness Account and Analysis of a Circles-Effect Event at Westbury' by G. T. Meaden, p 265
New Scientist No 1680, 2 September 1989, p 12
Time 11 September 1989, p 32
Sydney Morning Herald 4 January 1990, p 1
New Scientist No 1722, 23 June 1990, pp 25–27
New Scientist No 1723, 30 June 1990, p 8
Sydney Morning Herald 26 July 1990, p 1

Flying Cow

Hey diddle diddle, the cat and the fiddle, the hoatzin flew over the Amazon . . .

*t*he hoatzin (pronounced wat-sin) is a bird of the Amazon rainforest. It looks like a bird but gastronomically speaking it's a cow. And just to confuse things a bit more, there's a bit of monkey and a bit of fish mixed up into this mongrel.

The fully grown hoatzin bird has huge blood-red eyes glaring out of a bright blue face, with a fierce mohawk cut on top. But it's a small bird weighing about 0.75 kilograms, about the size of a pigeon. It hangs around in the Amazon rainforest, ranging from Guiana to Brazil, and it has an unusual BO problem. The hoatzin smells like cow manure so badly that the locals call it the 'stink bird'! It eats practically nothing except green leaves, and it is the only known bird that uses the same system of breaking down food that cows, sheep and deer use.

Grazing animals have a special gut that turns coarse fibrous plants into a high quality meal. But they don't do it alone. They have squillions of tame bacteria living in their gut to help them break down, or ferment, otherwise indigestible feed. These bacteria turn fibrous cellulose into simple fatty acids that are easily absorbed by the animal. They also make high-quality proteins, and all the vitamins the animal needs.

It turns out that these bacteria also live in the hoatzin bird, in a large swelling called the crop, half way down its food pipe. The crop, this special digestion breakdown box, is a combination of a pepper grinder and a compost heap—it has a mechanical function and a biochemical function. It has a very muscular and deeply corrugated wall that grinds the leaves up. Cows have a similar organ. And just like in cows, it is full of friendly bacteria that break down green leaves into important energy foods.

Now if you want to have friendly bacteria living inside you, you can't boot them out after every meal. An international team from Venezuela, the USA and Scotland found out all about the crazy guts of the hoatzin by feeding it tiny plastic beads. They found that if the plastic beads had a volume of one cubic millimetre they would pass through in 20 hours, but if they were three times bigger, they would take 43 hours to pop out as manure-smelling plastic plop. This very long time for hanging onto its feed is a record in the bird kingdom. It means that lots of bacteria can set up house inside the hoatzin bird, and not have to worry about being flushed out.

In fact this front end of the digestive

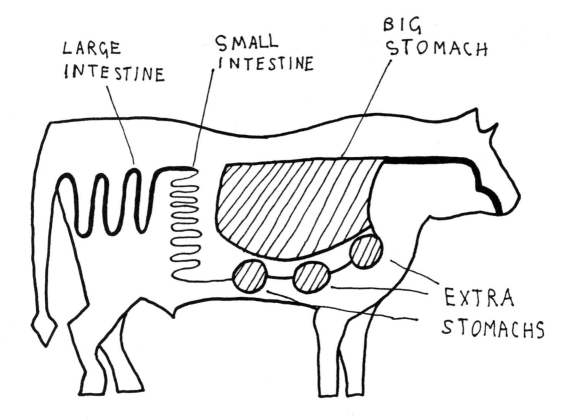

system is so important, that just the crop and oesophagus make up about three-quarters of the total weight of the gut. And the food inside this front end can account for 20 per cent of the total weight of the bird.

Now the advantage of having the gut of a cow is that even if the hoatzin bird can't find a worm, it can always find a leaf or two to nibble. The disadvantage is that the crop takes up so much room inside the body of the bird, that the breast bone is very small. This means that there's not much room to attach the flying muscles, and so the hoatzin is a weak flyer that never goes hungry.

They're so bad at flying that the baby hoatzins take about 65 days before they can lurch through the sky. This makes them fair game for any passing predator. But the baby hoatzins have fully operational claws on the first and second fingers of their wings. They can't fly, but they *can* get away from a killer by swinging with their wings from branch to branch, just like a monkey. And if things get really bad, they have been seen diving into rivers and swimming underwater like a fish.

So everything balances out in nature. You might have all the goodies and be able to eat leaves like a horse, fly like a bird, swing like a monkey, and swim like a fish, and never go hungry—but you'll always stink like cow poop.

REFERENCES

Science Vol 245, 15 September 1989 'Foregut Fermentation in the Hoatzin, A Neotropical Leaf-Eating Bird' by Alejandro Grajol, Stuart D. Strahl, Rodrigo Parra, Maria Gloria Dominguez, and Alfredo Neher, pp 1236–1238
New Scientist No 1684, 30 September 1989, p 15
Scientific American December 1989, p 18

Garlic is Good

*f*or a long time, people have known that garlic and onions are good for you.

Three and a half thousand years ago, the ancient Egyptian doctors wrote one of the very first medical text books, the *Codex Ebers*, on papyrus. It had over 800 drug prescriptions, and 22 of them had garlic as their main ingredient for treating problems such as headache, heart disease, worms and cancer. In fact, the Egyptian Pharaohs took carvings of garlic and onions into their tombs.

The ancient Greeks such as Aristotle, Hippocrates and Aristophanes used garlic as a medicine. But on the other hand, the ancient Greeks also stopped people who ate onions and garlic from attending worship at the Temple of Cybele.

The ancient Romans also loved garlic and onions. Pliny the Elder used them both in medicines, while around AD 100, Dioscorides, the chief physician to the Roman military, gave garlic to rid the bowels of worms.

In India, they used garlic as an antiseptic to clean wounds. And in China, onion tea has been used for thousands of years to cure dysentery, fevers and headache. In France, onions and garlic were given to horses to break up blood clots in the legs. In World War I, garlic-soaked bandages used on the wounds of British troops saved thousands of lives.

If you put chopped garlic cloves in a saucer on a window sill, it will keep the flies away. You can crush garlic and use it to make a non-toxic spray to keep away aphids, cabbage moths and caterpillars. And of course we all know what it does to vampires. But only recently has medical science realised that onions and garlic are worth a second look.

Onions are not yet as famous as garlic, but they might be able to stop an asthma attack. Depending on how you treat them, onions make a whole bunch of different chemicals. If you just cut a fresh onion, you get the chemical that makes your eyes water. But luckily this chemical is weakened by cooling, so if you don't want to cry when you cut onions, you can prechill them in the fridge or the freezer. This chemical is also soluble in water, so you might try cutting your onions under a running tap. If you fry onions, you get a different sweeter smelling chemical. But if you slice and chop the onion, you'll make another chemical that stops asthma attacks in guineapigs. This brand-new experimental chemical is still being tested, but it might be useful for humans. If you are an asthmatic, and can feel an attack coming on, chopping onions very fine might work, if you have run out of your standard anti-asthma drugs.

Garlic juice will stop the growth of more than 20 different kinds of bacteria and 60 different types of yeast and fungus. It's so powerful, that even if you dilute the garlic

juice to one part in 125 000, it will still stop bacteria such as the golden staph, *Staphylococcus*, *Streptococcus* which can cause rheumatic heart disease, *Vibrio* which can cause cholera, and *Bacillus* which can cause typhus, dysentery and food poisoning. Garlic won't kill these bacteria. But it will stop them from growing, and that will give your body time to mount its own defence.

One active anti-bacterial agent in garlic juice is a sulphur-containing chemical called allicin. Allicin is more powerful than penicillin when used for *Bacillus typhosus*. Allicin is also an anti-fungal agent. Many herbalists use it against thrush, which is a common fungal infection of the vagina. But the trouble is that allicin also helps cause that famous garlic odour.

This chemical, allicin, is not inside garlic, and appears only after the garlic bulb has been crushed or cut. Allicin is not found in the pills, oils and garlic extracts that you buy from your local health food shop. Its half-life is so short that most of it is gone within a few hours. So if you want to take garlic for its full antibiotic or anti-fungal properties, you can't just swallow a bulb of garlic whole. Either you crush it on the bread board, or you munch it in the mouth, or best of all you crunch and munch. Tough about the garlic smell though.

But garlic has yet another medical property. It stops the blood from clotting. Until now, one of the best anti-clotting chemicals was aspirin—the stuff you get from the bark of the willow tree. Aspirin is used by people who have had one ministroke and don't want to get a major stroke. It's also used by people who have had artifical valves implanted in their heart. The trouble with the artificial valves is that clots can form on them. Sometimes these clots can go to the brain, killing a part of the brain and causing a stroke.

Eric Block, a professor of chemistry at the State University of New York in Albany, has found a chemical in garlic that is better at stopping clots than aspirin. This chemical is called ajoene. It's being tested right now, and if successful, it could be a replacement for aspirin. Ajoene forms by condensing from allicin—it has never been found in any of these health food garlic extracts.

But garlic can do more. If you eat up to 10 cloves of garlic a day, it can help lower the levels of cholesterol and triglyceride in your blood. And a Chinese study in the province of Shandong has found that the more onions and garlic that the locals eat, the less stomach cancer they get.

So if you want the full anti-bacterial, anti-fungal, anti-clotting and anti-cancer effects of the garlic, you'll have to chomp your way through a bunch of fresh cloves of garlic and everyone will know that you've had the breath of life.

REFERENCES

Scientific American March 1985 'The Chemistry of Garlic and Onions' by Eric Block, pp 94–99
Australian Doctor Weekly 19 June 1987, p 47
Science Digest June 1988, p 92
New Scientist No 1684, 30 September 1989, p 16

The Great Attractor

*O*ut near the Southern Cross there's something sucking us all in—and until recently, we didn't even know it was there. The astronomers call it, rather dramatically, the Great Attractor.

Since the 1960s, scientists have believed in the Big Bang. This theory says that about 15 billion years ago, everything there was in the entire universe was contained inside a primeval fireball smaller than a cricket ball. Nobody knows how long this cosmic cricket ball was hanging around for, or what there was before it existed but suddenly the mysterious Big Bang happened, the universe started expanding, and the temperature went straight through the roof into the hundreds of millions of degrees. The naked throbbing energy cooled down into quarks, which cooled down into protons and neutrons. A few minutes later, the protons and neutrons joined up to make the cores of the lighter atoms, and then, 300000 years after the Big Bang, atoms formed, and that's today's explanation of how the universe began.

We have three things left over from that colossal explosion, 15 billion years ago. First, everything in the universe is still expanding away from everything else. Second, there is everything in the universe. And the third thing left to us from that big explosion is that

the enormous temperature of squillions of degrees has now cooled down to a temperature of only 2.7 degrees above absolute zero. The astronomers like to round things off, and call it the '3-degree radiation'. Almost everywhere you look in the sky, you see this almost perfectly even background level of radiation, the 3-degree radiation. They believe that this is the afterglow of the Big Bang.

It's very even almost everywhere, except in a small part of the sky close to the Southern Cross. In that direction, the 3-degree radiation is nearly four-thousandths of a degree hotter than everywhere else. This tiny increase in temperature might be minuscule, but it is still 80 times bigger than other bumps measured anywhere else in the universe. This was the first hint of the existence of the Great Attractor in the sky that is sucking us in.

So Allen Dressler in Washington at the Carnegie Institution, Sandra Faber in Santa Cruz at the University of California and five other astronomers started looking near the Southern Cross. The 'seven samurai', as they soon came to be called, started measuring the speed at which galaxies were moving. Now there is a general outward expansion of everything in the universe left over from the Big Bang, but once they had subtracted away this outward expansion, they discovered a very curious thing. There seemed to be an enormous gravitational field about 150 million light years away from our galaxy, roughly in the direction of the Southern Cross. (The stars of the Southern Cross are actually quite close, just a few hundred light years away.) They called it the Great Attractor. Thousands of galaxies, both on this side and the other side of this mysterious Great Attractor, were being dragged into it at speeds up to 1000 kilometres per second. It's like a huge plughole in the universe, sucking in not just galaxies, not even just clusters of galaxies, but clusters of clusters of galaxies. And it's *not* a black hole, because the galaxies at the Great Attractor itself are not disappearing, they just seem to be sitting there.

Now the basic unit of chemistry is the atom, and the basic unit of astronomy is not the planet or star, it's the supercluster. Our planet, earth, is one of at least nine planets going around the sun. The sun is one of at least 100 billion stars in our galaxy, which is called the Milky Way. The Milky Way is one of 20 galaxies in the Local Group. And the Local Group is one of many clusters of galaxies that make up the Local Supercluster, which has about 4000 galaxies in it. The supercluster is the smallest unit that astronomers use when they think about the large scale structure of the universe.

In fact, the universe seems to be made up of a bunch of bubbles, just like in a bubble bath. It's early days yet, but it seems that in the universe, all of the superclusters make up the walls of each individual bubble, and in the centre of each bubble there are vast amounts of empty space.

It's a bit hard to talk about the edge of the universe, but we can talk about the most distant parts of the universe that we know about. The Great Attractor is about one-hundredth of the way to the edge of the universe, and it seems to be some sort of huge gravitational mass inside one of the walls of the bubbles that make up our bubble bath universe.

But what exactly is the Great Attractor made of? Well Dressler and the rest of the seven samurai have worked out that the Great Attractor is about 300 million light years across, making it the largest single structure ever discovered in the universe. And they reckon that it must weigh as much as tens of thousands of galaxies. But the funny thing is they can't see anything where the Great Attractor should be.

Don Mathewson, professor of astronomy at the Australian National University, reckons that the Great Attractor is a cosmic string. Cosmic strings are one of the latest concepts that astronomers use to explain mysterious things like the hidden Great Attractor.

Cosmic strings are impossibly heavy, one-dimensional, thin tubes of ancient high-energy vacuum. They're incredibly skinny, one-millionth of one-millionth of one-millionth of one-millionth of one-millionth of a centimetre. Because they're so skinny, they're invisible. But to make up for being so skinny, they're very heavy and each centimetre weighs about 100 million billion tonnes. They're so heavy because they have stuff from the Big Bang cricket ball trapped inside them. It's not the sort of string you use to tie up a parcel.

According to the theoretical physicists (they're the ones with the big heads and the sharp pencils) cosmic strings were made in the very early days soon after the Big Bang. Cosmic strings don't have any ends. So either they're infinitely long, or else they're tied up

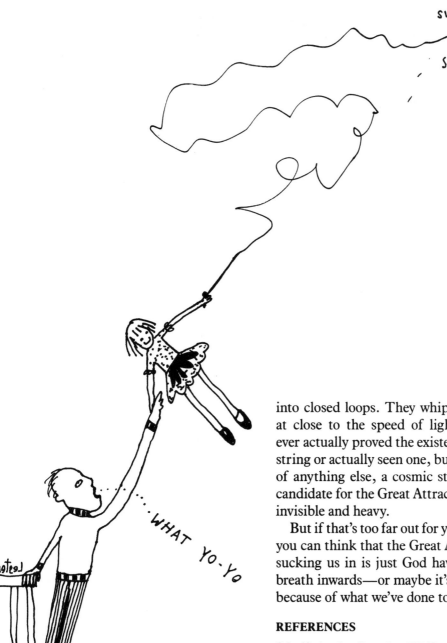

SUCK

SUCK

WHAT YO-YO

lester

THANKS DAD! THIS COSMIC YO-YO IS WILD!

into closed loops. They whip through space at close to the speed of light. No one has ever actually proved the existence of a cosmic string or actually seen one, but in the absence of anything else, a cosmic string is a prime candidate for the Great Attractor, because it's invisible and heavy.

But if that's too far out for you to believe in, you can think that the Great Attractor that is sucking us in is just God having a big deep breath inwards—or maybe it's a cosmic sigh, because of what we've done to our planet.

REFERENCES

Scientific American December 1987 'Cosmic Strings' by Alexander Vilenkin, pp 52–60
The Sciences September/October 1989 'In the Grip of the Great Attractor' by Allen Dressler, pp 28–34
New Scientist No 1684, 30 September 1989, p 15
Discover November 1989, pp 20–22
New Scientist No 1702, 3 February 1990, p 14
Scientific American January 1990 'The Cosmic Background Explorer' by Samuel Gulkis, Philip M. Lubin, Stephan S. Meyer and Robert F. Silverberg, pp 122–129

Halley's Comet

*h*alley's Comet was first noticed and written about by our distant ancestors 120 generations ago. It was only eight generations ago that we noticed that it makes a regular visit every 76 years. And in March 1986, we sent out a welcoming party of five spacecraft equipped with our latest technology.

Two Japanese spacecraft took the wide-angle shots at distances of some 100 000 kilometres, and 10 million kilometres, while two Soviet VeGa spacecraft took a closer look from around 9000 kilometres. But the European Giotto spacecraft went for the ultra-close-up, at a distance of only 600 kilometres, zipping past it at 70 kilometres per second. An astounding amount of information was recorded, and astronomers have been analysing the results ever since. We now think Halley's Comet could possibly be an alien.

The popular theory is that before most comets fly through our part of the solar system, they hang around together a long way out from the sun, somewhere between half a light year and two light years away. It's cold and dark out there—the temperature is only 4 degrees above absolute zero (−269°C), and the sun is just another bright star in the sky. Millions of lumps of dirty ice circle slowly around this distant star.

But every now and then there's a shake up in this cloud of frozen future comets—a star

might pass close by, or some of the lumps of dirty ice might run into each other. Somehow, one lump of dirty ice is shaken out of orbit, and begins its million-year fall towards the distant sun. As it comes near the inner planets, the heat from the now-close sun makes some of the ice evaporate into water vapour. The dirty snowball begins to rock and roll, and creak and tumble, jets of gas and dust start spurting, and soon a large tail stretches across our night sky.

We now know that the core of Halley's Comet, the solid bit, is about 16 kilometres long by 7.5 kilometres across and 7.5 kilometres wide, about the size of a small tropical island. It has the shape of a peanut shell, and a total volume of 500 cubic kilometres. But it weighs only 100 billion tonnes, so it's about one-fifth as dense as ice. If Halley's Comet fell into an ocean, and survived the fall, it would float.

On the outside of Halley's Comet, there's a hard-baked crust, that has been burnt by going so close to the sun. This heat-resistant material is very dark, as black as soot. It's

probably made up of organic compounds. In fact Halley's Comet seems to be covered with the same chemicals you would get if you got human beings (or pork pies), freeze-dried them, crumbled them up and then exposed them to the heat of the sun.

Underneath the protective crust, there's a layer of half-cooked material made up of rocky dust, carbon and water-ice probably fluffed up into some sort of honeycomb shape. Deep inside Halley's Comet we would expect to find gases that have been frozen into ice and will evaporate once they get warm.

Halley's Comet spins around once every 53 hours or so. When the armada of spacecraft visited it, it was about 150 million kilometres from the sun, and at that time its temperature was a chilly −70°C on the night side, and an almost boiling 90°C on the day side.

Ten per cent of the side facing the sun is covered by half a dozen distinct isolated jets, spewing out the dust and the gases that make up the tail. And of course, as the black peanut spins, the jets on the sunlit side switch off when they go into the dark, and new jets start spurting as the dark side comes into the light. There are about 20 tonnes of gas shooting out each second, and about 5 tonnes of dust.

The gas is made up of about 80 per cent water, with about 10 per cent carbon monoxide, 3 per cent carbon dioxide, 2 per cent methane, less than 1.5 per cent ammonia, about 0.1 per cent hydrocyanic acid, and lots of other chemicals. These chemicals are made of the same elements as you and me.

There are three different types of dust squirting out at 5 tonnes per second. The first dust is a rather rocky mix of magnesium, silicon, calcium and iron like planet earth. The second type of dust is also rocky, but has more sulphur, carbon and nitrogen, like lava. But the big surprise is the third kind of dust, which is made up of mostly carbon, hydrogen, oxygen and nitrogen. These are the chemicals of life on earth.

On this last visit, Halley's Comet lost only 100 million tonnes or just a thousandth of its total weight. So we can expect a few hundred more happy returns, hopefully with more impressive fireworks than we saw this time.

But the last big surprise is that Halley's Comet is probably an alien. There are a few different types of carbon, and the most common are called carbon 12 and 13. Everywhere in our solar system, there are about 90 times as many carbon 12 atoms as carbon 13 atoms. But Halley's Comet has a ratio of not 90, but 65 which is the same proportion we've measured in the interstellar gas in the space between the stars.

Halley's Comet could be an adopted orphan from somewhere outside our solar system. After all it's one of the very few objects that goes around the sun in the opposite direction to us and all the other planets.

So now we know that Halley's is made of the chemicals of life as we know it but we also think that it comes from beyond our solar system. This could mean that the elements of life are widespread throughout the Milky Way, or that we are not alone, that life is everywhere in the universe, perhaps even life as we know it.

REFERENCES

Astronomy September 1986 'What Have We Learnt From Comet Halley' by Richard Berry and Richard Talcott, pp 6–22
Astronomy June 1987 'Search for the Primitive' by Richard Berry, pp 6–22
Scientific American September 1988 'A Close Look at Halley's Comet' by Hans Balsiger, Hugo Fechtig and Johannes Geiss, pp 62–69
New Scientist No 1660, 15 April 1989, p 31
Discover August 1989, p 26

Hot Anti-World

*t*here is a very strange and exotic world that is an upside-down mirror image of planet earth. It is hotter than the surface of the sun, closer than Los Angeles, and 10 times as weird. It has anti-oceans and anti-mountains, mirror images of our oceans and mountains, and rainstorms of iron filings.

This bizarre world is straight under our feet. It's a ball of molten iron and nickel at the centre of the earth, roughly the size of the moon.

It's absurd and illogical, but we know more about the distant stars than we do about the ground beneath our feet. We can't see through rock, so the only way we can get a glimpse of this inner world is from the shock waves from earthquakes. When the shock waves travel through heavy dense rock they speed up, and in lighter rock, they slow down. Over the last century, 3000 monitoring seismic stations scattered around the planet have collected the records of 25 000 earthquakes. By analysing how long it took shock waves from earthquakes to travel to different seismic stations around our planet, the geophysicists slowly built up a very rough picture of our inner world. By the 1940s, they had a model of our planet that looked a little bit like an onion—with three main layers.

The first and thinnest layer is called the crust. It's only about 25 kilometres thick, roughly three times the height of Mount Everest. Compared to the whole planet, it's like a postage stamp stuck onto a soccer ball. Inside this very thin crust are all the rocks and sediments that we know about, the pockets of oil and gas that we drill for, and the volcanoes. The world's deepest hole, in the USSR, reaches only about 12 kilometres into the crust, and that's only half way.

The next layer is the mantle, which stretches from the crust to 3000 kilometres below the surface. In the mantle are many different layers of very hot rock. The mantle is stirred by currents, flowing like boiling water, but in super-slow motion, rolling and tumbling like a very thick glue or molasses.

And right at the centre of our planet is the core. It's about 6500 kilometres across, roughly the size of Mars. It's metallic, mostly iron and nickel. And following on with our onion model, the core is divided into two main layers. The outer layer is molten liquid—one theory is that it circulates, causing the earth's magnetic field. And right

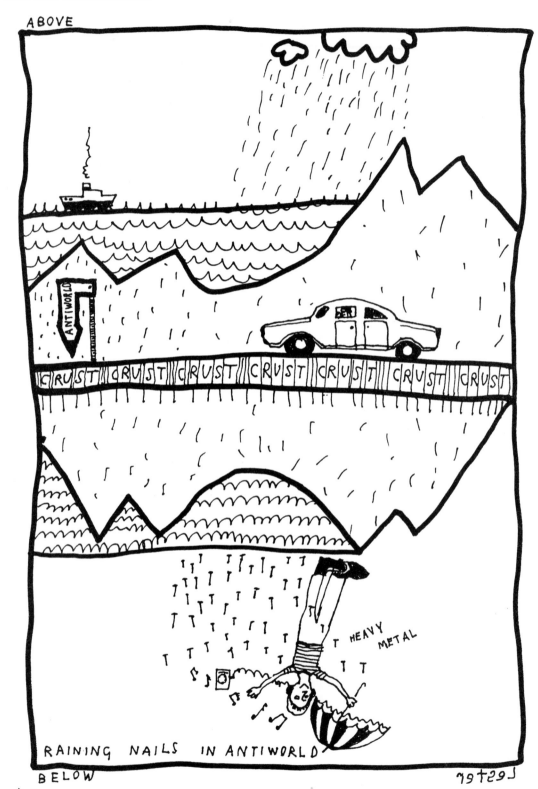

at the very centre of the earth, about 1200 kilometres across or the size of our moon is a solid core of iron and nickel.

A group of geophysicists at Harvard University, led by Adam Dziewonski, wanted to find out more about the surface of this solid ball of iron and nickel. They borrowed a technique from medicine. Doctors use CT scans to look inside the body. Thousands of X-rays are squirted through your body from all angles, and are picked up by a bunch of receivers. A three-dimensional picture can be built up from this information. To make a planet-sized CT-scan machine, you'd have to set off huge explosions to pump shock waves big enough to go through the planet. Nuclear weapons are a very messy way of doing this, but luckily for the geophysicists, there are lots of natural earthquakes.

So the geophysicists went back to the records of 25 000 earthquakes, some of them a century old, fed them into a CT-scan computer, and came up with an astonishing new picture of the core. The surface of the solid core is not smooth. It is dimpled with valleys some 5 kilometres deep, and spiked with mountains taller than Everest, some 10 kilometres high. Where we have mountains and continents sticking up towards the sky, thousands of kilometres directly underneath, there are anti-mountains pointing down towards the centre of the earth. And where we have oceans dipping down into the crust of the earth, on the core are anti-oceans bulging up. It's a weird and curious upside-down shrunken copy of our world.

Other scientists found the solid metal core is at a temperature of 6600°C—that's hotter than the surface of the sun, but it's closer than Los Angeles! And this strange super-hot anti-world even has its own bizarre iron rain. The equator of the core is slightly hotter than its poles. So the iron boils off from the equator and travels up towards the poles, where it condenses and falls down again as an iron rain—a weird geological weather system.

I wonder if on this weird anti-world, there is anti-life, and if so do they listen to hard-core rock?

REFERENCES

Discover November 1987, pp 86–93
Time June 1988, p 61
Popular Science November 1988, pp 76–81, 119–120
Scientific American April 1988, pp 16–17

Kangaroo

*t*he female red kangaroo (she's actually bluish; only the male is red) has three vaginas and two different flavours of breast milk on tap. As for the males, most of them die virgins. And if they do get lucky, they die young.

The kangaroo is a unique creature. No other animal its size can hop, and growing up in Australia's harsh climate has meant that the lazy roo has some unusual optional extras.

When kangaroos hop, they get better fuel ecomony than any other creature. Hopping at full speed they use practically the same amount of energy and air as they use at slow speed. They use a couple of special features to be so efficient. The tendons in their back legs are very springy, like pogo sticks: they store and give back about 90 per cent of the energy needed for a hop. After spending some energy to get moving with its first hop, a kangaroo can then bounce along doing only one-tenth as much work per hop. Human leg tendons are much less springy: they give back only 30 per cent of the energy.

Roos also don't get out of breath when they go fast. When a kangaroo springs up the pouch and the beergut lag behind a bit—and pull the lungs down. So air is automatically sucked into the lungs. Roos don't have to use their chest muscles when hopping. And when Skippy lands, the gut squashes the lungs, and pushes the air out. The gut acts like a piston to force air in and out of the lungs. So kangaroos take one efficient synchronised breath every hop. We humans are not so

efficient and have to take two strides for each breath.

Now Skippy might be a rugged highly efficient performance machine, but lurking behind his sleek appearance is a promiscuous and competitive sex life. In every tribe of kangaroos, there is usually one big feller, who gets 80 per cent of the successful bonks. He's called an alpha male. The junior males all have to share the remaining 20 per cent, usually behind his back, while Mr Big is getting it off with the rest of his harem.

Before the junior males get a chance to become Mr Big they have to be tough enough to beat him in a fight. This means that they have to be about 10 years old and weigh about 85 kilograms. But any new Mr Big will not last long. Twelve months is tops, even if he doesn't get beaten in a fight by some young punk. A Mr Big in the kangaroo world has 10 years of pent-up frustration crammed into one year of frantic bonking—and maybe it's too much fun. Within 12 months, for some unknown reason, he gets senile, blind, paralysed and dies.

While the kangaroo boys live fast and frenzied sex lives, the red kangaroo women have bizarre naughty bits. For example, look at the uterus. This is the place where baby

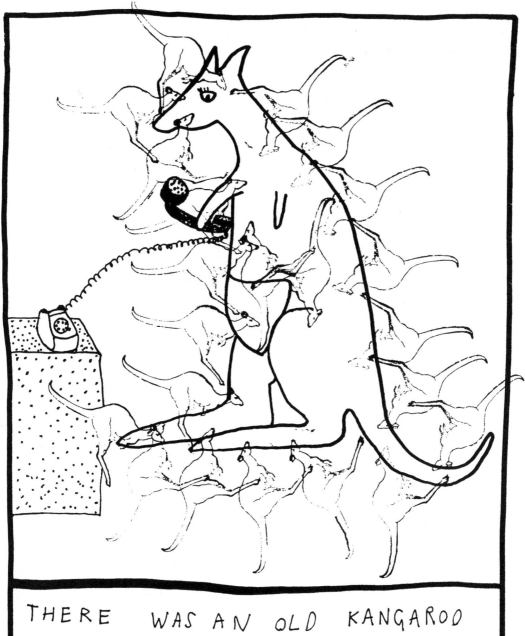

THERE   WAS AN OLD   KANGAROO
. . . .
SHE HAD SO MANY   CHILDREN
SHE DID'NT KNOW WHAT TO DO!!!

joey forms. Women have only one uterus—but female red kangaroos have two! And while women have only one vagina, female red kangaroos have three. The two lateral or side vaginas carry the sperm up to the egg, and when the time is right, the middle vagina carries joey to the outside world. But there's only one opening on the skin, because all three vaginas join up just inside. It's a strange plumbing system, but millions of red kangaroos say that it works just fine.

A roo preganacy lasts for 33 days. When the joey is born, bright red and weighing less than a gram, it uses its giant nose and the claws on its front legs to struggle up to that place where the milk is, the pouch. It has absolutely no hind legs. It hooks onto the first nipple it finds. Joey has a strong large muscular tongue, with which it can suck powerfully. Sucking on the nipple releases a clear thin baby formula, and also makes the end of the nipple swell up inside the joey's mouth. In fact, after just one day, joey can't spit out the nipple even if it wants to.

When joey number one is all of two days old, mum can get pregnant again. The fertilised egg lands in the other uterus, but this time, there's a difference. The egg, 0.25 millimetre in size and only 100 cells, does not get bigger. It gets put into suspended animation and can stay in this biological cold storage for more than 200 days.

Five months later, joey number one finally pokes its head out into the open air for the first time and begins to eat grass while its body is still in the pouch. And six weeks later, it has its second birth and finally leaves the pouch—so just like some Christians, all kangaroos are born again. As the hopping joey drinks less milk, joey number two in cold storage thaws out, starts to grow, and 33 days later, is born.

Mummy kangaroo can have three joeys at the same time: joey number one on nipple number one, drinking a rich thick milk, but hopping around and eating some grass as well; joey number two stuck inside the pouch on nipple number two, and sucking a thin clear baby formula; and joey number three inside one of the two uteruses in cold storage.

Scientists don't know how a fertilised egg can be kept in cold storage and not get older. Maybe the kangaroo also has the secret to eternal youth.

REFERENCES

Scientific American August 1977 'Kangaroos' by T. J. Dawson, pp 78–89
Alumni Papers UNSW, Winter 1987, pp 13, 14
Sydney Morning Herald 14 August 1987, p 3
University of Sydney Gazette September 1987, p 3
The Australian Way November 1987, pp 51–53

Killer Bees with Chips

*k*iller bees are invading North America. They're very very nasty and it's all our fault. We humans created killer bees and now we're paying the price.

In 1957 African queen bees were accidentally released from a laboratory in Brazil. These African queen bees bred with the local bees and formed a new species, a hybrid called Africanised queen bees. But now they're called killer bees: they are bad tempered and aggressive and have actually killed people and animals with their relentless stinging. After a bee stings an animal an 'alarm odour', which smells like overripe bananas, is left behind. Other bees attack the animal that is giving off this smell. The Africanised killer bees are more likely to attack.

These hybrids have been moving north from Brazil. Right now they're moving at about 500 kilometres per year, roughly the distance from Sydney to the Victorian border, and have already crossed the Mexican border. Just like the Mexican wetbacks (illegal immigrants), they have struck at the heart and mind of the USA. The Americans are worried that killer bees could be a threat to their bee industry, which is worth $25 billion a year.

Killer bees look similar to ordinary bees, so how can you recognise them? Killer bees beat their wings faster than their European cousins. So Kelly Falter, a nuclear engineer at the Oak Ridge National Laboratory in Tennessee, developed a special hand-held intelligent microphone that uses noise analysis circuitry to measure the frequency at which the bees' wings beat. You simply point it at a suspicious-looking bee and it flashes a red light for killer bees, and a green light for ordinary bees.

But the engineers have gone even further and have put flashing beacons on the killer bees. So now they know exactly where the killer bees are, and how they spend their day.

If we want to win the war against the killer bees, we need to know more about them, so that we can pull a few dirty biological tricks and avoid using pesticides. We want to know what and when they eat, and we want to know about their sex life. One thing that we do know is that killer bees don't have a lot of sex, which could be why they're so irritable. But when they get it, it's probably really good. The queen of the killer bees has only one mating flight in an entire lifetime. After that, the bee experts don't know what goes on. This queen takes off, and vanishes over the nearest hill at 22 kilometres per hour. It is impossible to see where she goes because she is so small and fast.

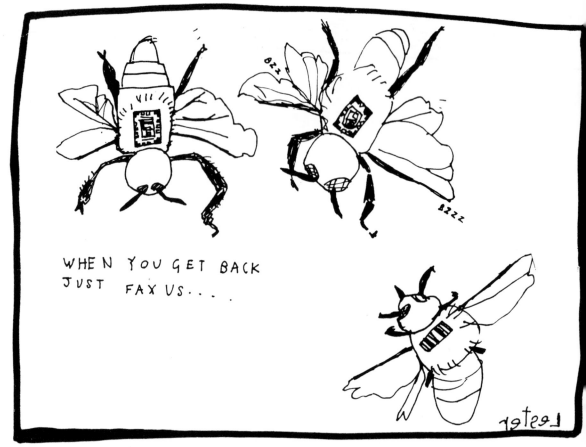

WHEN YOU GET BACK JUST FAX US. . . .

But you can find a bee if she has a homing beacon on her. So Falter, working with a team, built a tiny solar-powered transmitter on a single minuscule silicon chip. It weighs about 35 milligrams, as much as a grain of sand—that's about half the load that your average 88-milligram bee can carry. The silicon chip has nine tiny solar cells that catch the photons in sunlight and turn them into electrons to store the energy. They slowly trickle these electrons into a capacitor, which is like a rechargeable battery—it stores energy. And once enough electrical energy has been stored up, the capacitor then dumps this energy into a light emitting diode (LED). The LED turns this electrical energy into invisible infra-red light that can be picked up by receivers up to 2 kilometres away. So this means that bee experts can track bees, in the same way that biologists have been tracking wolves and whales and seals with tiny radio transmitters.

The tricky bit is holding the bee perfectly still, while you glue the chip onto its back. And if they're lucky, and glue the chip in exactly the right position, then the bee can still fly.

The bee scientists want to get clues that will help them stop the spread of these Africanised killer bees, without having to use pesticides. And they will do it because the bee will have a chip on its shoulder. The bugs will be wired for sound—they'll be bugged.

REFERENCES

New Scientist No 1631, 22 September 1988, p 37
Popular Mechanics November 1988, p 13
Scientific American January 1989, pp 14, 15
Popular Science February 1989, p 11

Lightning

*l*ightning smashes into the earth 100 times per second, generating more than 4 billion megawatts of continuous power—about 100 000 times more electrical power than all of Australia's power stations generate.

But lightning also protects us from charged radiation from deep space. The ground has a negative charge, while the earth's upper atmosphere has a positive charge. When a positive particle zooms in from the depths of the universe, it is repelled away from earth by the positively charged atmosphere. The 100 lightning bolts each second constantly rebuild the charge barrier, this natural 'Star Wars' shield, and protect us.

Your average lightning bolt lasts for about one-fifth of a second. The whole thing starts with a big dark storm cloud. Somehow—the lightning scientists still don't know how—the cloud gets a huge positive charge on the side closest to the ground. But remember, the ground has a negative charge. Opposites attract, and positive charge loves to merge with negative charge. Just like sexual stress, electrical stress has to be relieved.

Lightning bolts happen in two stages. In the first stage, a relatively slow small skinny almost invisible bolt of lighting—called the down stroke—creeps down from a cloud to the ground in steps between 1 and 200 metres long, blasting a pathway through the air at 150 kilometres per second. When the down stroke is about 15 metres from the ground an enormous charge rushes out of the ground back up to the cloud. This is the second stage—the return stroke—which is what you see and hear. It's only a few centimetres in diameter, but it carries a million times more energy than the down stroke. The return stroke rockets back up to the cloud at about 150 000 kilometres per second, or half the speed of light. It travels up inside the hole that was blasted in the atmosphere by the down stroke. The temperature inside the return stroke reaches 30 000°C—that's six times hotter than the surface of the sun!

Lightning kills about 100 people each year in the USA. The outdoor life is dangerous. If you're in a thunderstorm, and you can feel your hair standing on end, or you can hear buzzing noises, you're in the powerful electrical field of the strike zone. Get out quickly. The first rule is go, go, go.

Now even if lightning misses you and hits the ground nearby, it still sets up enormous electrical fields. If your feet are far apart, there can be a potential of several thousand volts between them, which is enough to kill you. The bigger the distance, the bigger the voltage. So if you lie on the ground, the voltage between your head and your feet could be 20 000 volts. So crouch down into a tight little ball, and keep your feet together.

In fact if you really want to be cool, you should squat on one foot, so electricity can't come in through one foot, through your heart and out through the other foot. So the second rule is—get down, and make like a one-legged rock.

If your clothes are totally wet, they will carry electricity better than your human flesh. So don't just get down, get wet. You may get a few skin burns, but the electricity will miss your heart and soul. If your skin is wet you can even have the clothes burnt off your body, which is left totally unharmed, but pretty shocked. The lightning energy heats up the water on your skin instantaneously and turns it into steam which expands and blows your clothing off your body, leaving it unharmed. Often you can be knocked unconscious, or even have a temporary mini stroke. Ten French and German tourists who hid under a bush during a thunderstorm in Trento, Italy, in 1970 had their clothes stripped off them by a lightning bolt—at least, that's their story.

If a tree is totally wet, it's fairly safe to get under it. A tree is God's natural lightning conductor. The electricity will hit the top of the tree, and run down through the layer of water on the leaves and the bark, and then into the ground. But if the bark is dry, the electricity will get to the ground either by jumping off the bark onto you or by tunnelling into the tree trunk. The sap in the trunk can boil instantaneously, and explode with the force of 250 kilograms of TNT. In Darwin in January 1987, 10 joggers were thrown off their feet when lightning shattered a nearby tree. One man actually stopped breathing, but he was revived by mouth to mouth resuscitation. In England in 1974, an 11-year-old girl died after her skull was cracked open by a flying piece of bark. So make sure your tree is all wet—one-quarter of all people killed by lightning were hiding under a tree. But if you don't feel lucky, avoid the tree, and just get wet.

Any metal outdoors in a lightning storm is dangerous. So get rid of your metal golf clubs, aluminium cricket bats and even your medallions. Even fishing rods are a good target.

Lightning rods are there to give a safe pathway for the electrical energy down to the ground. So a good lightning rod has a sharp point at the top end, which should be several metres taller than the house. The other end should be buried deep into the ground. The energy goes through the rod straight into the ground completely bypassing your house. In St Paul's Cathedral in London, the lightning rods are as thick as your arm. But during a huge storm in 1772 the iron bars glowed a dull red colour as they carried the lightning energy down to the ground. If only they had realised that electricity can make things glow, they could have invented the light bulb one hundred years before Edison.

Before lightning rods were common, and science replaced superstition, church-goers had to ring church bells during thunderstorms to appease the anger of God. But in France, in the 33 years between 1753 and 1786, a hundred bellringers died from lightning which struck the tall pointy bell tower and then ran down the wet rope. So, to stop this crazy human sacrifice, the French parliament had to make a law forbidding the ringing of bells during thunderstorms.

Telecom Australia had 326 reports of telephone-related lightning injuries in the eight-year period from 1977 to 1985. In a lightning storm, you should stay about one metre away from chimneys and electrical appliances, which should be switched off.

In fact, you should pull the plug out of the wall-socket. You certainly shouldn't use the telephone, because the lightning could get you.

Thunder is the noise that lightning makes as it blasts through the air. When a plane breaks the sound barrier, it moves at only 1 kilometre every 3 seconds. But lightning can move half a million times faster, so it makes a huge tearing noise when it pushes the air out of the way. You can use thunder to measure how far away a bolt of lightning is. The light from the bolt travels to your eyes at the speed of light, about 300 000 kilometres per second, which is virtually instantaneous. However the sound travels much more slowly, at only 0.3 kilometres per second. So as soon as you see the lightning flash, count the number of seconds until you hear the thunder. Then divide the number of seconds by three and that's the distance in kilometres away that the lightning blast was. So if you count 6 seconds, the lightning bolt is about 2 kilometres away.

Now there are a few reports of a crackling sound as soon as the witnesses see the lightning. Then, a few seconds later, they hear the thunder. Scientists think that the lightning causes voltages in things like household wiring or water pipes near to where they're standing. That takes care of 90 per cent of these reports. But sometimes the people were just standing out in the open near a wooden shed and still they heard those strange snapping, hissing, clicking or crackling noises. Once again the scientists have no explanation.

Sometimes you just can't avoid lightning.

Roy Sullivan of Virginia has been hit seven times! In 1942 he lost a nail from a big toe, in 1969 his eyebrows were burnt off, in 1970 his left shoulder was seared, in 1972 his hair was set alight, in 1973 it was burnt again before it had a chance to regrow, in 1976 his ankle was damaged and in 1977, his chest and stomach were burnt. He doesn't work as a park ranger any more. But he's such a natural conductor, that maybe he should run an orchestra, or be a bus conductor.

REFERENCES

The Medical Journal of Australia Vol 144, 23 June 1986, pp 673, 674, 706–709

New Scientist No 1592/1593, 24/31 December 1987, pp 64–67

Scientific American November 1988 'The Electrification of Thunderstorms' by Earle R. Williams, pp 48–64

Limping Metal Gods

*V*ulcan was the Roman god of fire and metalworkers. He was big and muscly like a blacksmith, but all the legends say that he walked with a limp. Industrial health workers now think that Vulcan got his limp from the first known industrial disease—arsenic poisoning.

The human race has been on the planet for three million years, but only recently have we been using metals. We started with copper, because in those days, you could find it lying around on the ground. It is one of the few metals that occurs naturally. The working of raw copper began about 11000 years ago in the Middle East, near Turkey. And only about 5500 years ago, in the early Bronze Age, smelting became widespread, and the early metalworkers refined a fairly pure copper.

But the trouble with copper was that it was too soft and didn't hold a sharp edge in battle. The early metalworkers soon found out that if you beat the copper with a hammer it would become a little harder and so hold its edge longer—but it still didn't make very good swords.

But then they discovered that if you added a small amount of arsenic to the copper while you were smelting it, the copper would become much harder. But they had to be careful; if they added more than 2.5 per cent arsenic, the copper became brittle—and a brittle sword could splinter just like glass. These ancient metalworkers were able to keep the level of arsenic around the 2 per cent mark—this was an incredible achievement 5500 years ago.

Even though it was an effective technique, the ancient metalworkers used it for only about 400 years—a very short time. Then abruptly, throughout the entire European and Middle Eastern world, the metalworkers swapped arsenic for tin. If you add tin to copper you get bronze, a much stronger and more versatile alloy. Bronze weapons then ruled the world for thousands of years.

The ancient metalworkers didn't have unions, but they did have their patron gods of metalwork. While the Romans worshipped Vulcan, the Greeks praised Hephistos, the Germans adored Wieland, the Scandinavians idolised Wolunder, and the Finns revered Ilmarinen. To show respect, these ancient

HE ALWAYS LIMPS AFTER A NIGHT ON THE POISON!!!!

metalworkers panel beated holy pictures of their particular patron god into copper, and every single one of these gods is shown limping!

Arsenic is a nasty poison. Criminals made it famous—a single dose of less than a quarter of a gram will kill most people. Pope Alexander VI, one of the horrible Borgias, poisoned himself accidentally by eating arsenic-laced food that was destined for his unwanted guests. Napoleon probably died of arsenic poisoning too. His hair was found to have 13 times the average arsenic content. Some scientists believe he picked it up from the green dye in his wallpaper: his house was very damp, and mould growing on the damp wallpaper turned the arsenic in the dye into another, very active form of arsenic.

But if arsenic doesn't kill you with a single large dose, many small doses can leave you with dermatitis, a hoarse voice, loss of weight and appetite, irritability, cancer and even a god-sized limp. Arsenic causes an inflammation of the nerves, especially those furthest away from the centre of the body, resulting in a burning sensation on the soles of your feet, or 'foot drop' where some of the leg muscles are paralysed so that you can't lift up the front of your foot. So it seems that as the ancient metalworkers were forging these new-generation sharp-edged swords, they were also breathing in fumes loaded with toxic arsenic.

Arsenic seems to be everywhere. Our bodies actually need a tiny trace of arsenic— the average human body has about 10 milligrams—while Vichy mineral water has 2 milligrams of sodium arsenate per litre. The first cure for syphilis and one of the earliest artificial drugs, Salvarsan, contained arsenic. If chicken and goats are fed on an arsenic-free diet, their growth is stunted. In fact, in America today, pigs and hens that are fed an arsenic tonic put on about 3 per cent extra weight, which more than covers the cost of the tonic. And when it comes time to kill the animals, the farmer gives them a few arsenic-free days to lose the poison out of their bodies. In those few days, the amount of arsenic in their meat falls to less than one part per million.

And maybe that's why they still keep chooks and pigs in tiny, cramped cages—so you can't see them limp, like the ancient gods of metalwork.

REFERENCES

New Scientist No 1488/1487, 19/26 December 1985, pp 10–14
British Journal of Industrial Medicine Vol 44, 1987 'Possible Toxic Metal Exposure of Prehistoric Bronze Workers' by M. Harper, pp 652–656
New Scientist No 1583, 22 October 1987, p 35

Micromachines

*t*he human race has had a long love affair with machines. The first machine ever used by humans was probably the pointed stick. It was used for digging and maybe stabbing. Machine technology then leapt ahead to the lever, to steam and electric engines, and then to microchips.

Now Californian engineers have built the world's smallest machines. To the naked eye, they look just like a speck of dust. But they are genuine machines, with tiny gears, pulleys, lenses, ball bearings, springs and rotors. And the engineers don't assemble them with microscopes, they grow them with chemicals.

These tiny machines, or micromachines, are made with the same techniques and materials that are used to make silicon microchips. The chips inside computers and digital watches have millions of separate electronic parts—transistors, capacitors and resistors. But they're not put into place one at a time—instead, they are grown in batches onto a sheet of very pure silicon, called polysilicon.

Normally, polysilicon is resistant to strong acids. But if you blast a powerful beam of electromagnetic energy—like X-rays—at polysilicon you change the very structure of the material. Those same acids will now dissolve the polysilicon in the blasted areas.

In a micromachine laboratory, the engineers start off by making a stencil of a mechanical part, like a gear. This stencil is actually a black and white silhouette of a gear, but is photographically reduced very small—as small as a bacterium, if you want. Then they lay it over the polysilicon. They spray the polysilicon with powerful beams of X-rays or electrons. The polysilicon around the stencil is hit by the energy beam, and it's changed. Then they pour the acid onto the sheet of polysilicon, and a tiny gear is left behind. But a single gear is as useful as a single shoe. So they put two gears on the same stencil, side by side, and eventually they grow all the different parts of a micromachine.

But how do you power a machine that's the size of a speck of dust? Right now, they're using motors that run on static electricity, but they're also testing hydraulic motors. In the future, micromachines will even run on the same energy that runs the human body, ATP. So you could implant a micromachine that would run as long as the person lived.

So far, they have made motors that are lighter than a fleck of talcum powder, and about 70 microns in diameter—that's the same thickness as a human hair. Inside the

AHH CHOO!!

motor are gears with notched teeth that are as small as the human red blood cell. They have made tiny axles thinner than a cobweb, and wheels that spin at 24000 revs per minute, and can spin up to that speed within half a second. One motor that weighs one-third of a gram can pull a whole packet of biscuits weighing 200 grams.

Already microsensors—the eyes, ears and nose of a micromachine—have been made. They can measure pressure, smell, temperature, sound and acceleration. In luxury cars they trigger air bags in crashes, and tell you when your tyre pressure is low.

At microsizes, silicon is as strong as steel, although in large chunks it's quite brittle. Silicon micromachines will work at temperatures up to 1400°C—that's robust enough to survive inside a car engine. The added advantage of using silicon is that you can make a microcomputer brain on the same slab as the machine, with no connecting wires. A micromachine with a brain and a body is a tiny robot.

There will be millions of uses for these machines, especially in medicine. Tiny excavating machines could be injected into the blood stream to find fatty deposits of cholesterol in the heart and nibble them away. Surgery could be done one cell at a time. Miniature chainsaws could be sent in to do delicate microsurgery in critical areas like the retina of the eye. They'd be like miniature SWAT teams inside your body, cruising through the blood vessels, exterminating the enemy.

The pills that you swallow in the future could have tiny machines that would slowly release an exact dose of a drug. Micromachines could even measure the glucose level in a diabetic, and dribble out the exact amount of insulin needed. And when the micromachines wore out, they'd be flushed out of the body through the kidneys.

Inside computers, if there is a break in an electrical circuit, a tiny kamikaze mini robot could go in there and find the fault. Then it could simply close the circuit by laying itself across the gap. It would stay there forever, as an intelligent, sleeping piece of wire. And you could use millions of these tiny robots to clean barnacles off the bottoms of ships, or clean the insides of obsolete nuclear reactors.

Already there is a shortage of engineers in the micromechanical field, as the military and multinationals such as General Motors, Ford and Honeywell pour money in. The smart money is tipping micromachines as the next big thing.

These machines are not only made with the same methods used to make silicon chips but they'll probably also end up being just as cheap. One day you'll get 10 microchainsaws for a cent.

But there is one big problem with these tiny machines. If you sneeze, or drop one on the carpet, you'll never find it again!

REFERENCES

Omni January 1986, pp 66–68, 104–108
New Scientist No 1628, 1 September 1988, p 44
Discover March 1989, pp 78–84
Popular Science March 1989, pp 88–92, 143
Time 20 November 1989, pp 60–61

New Telescopes

*W*here did the universe come from? Was there really a Big Bang, and how were galaxies made from its super-hot gases? How is matter scattered through the universe? What happens in newborn stars, and around the centres of galaxies?

Each of these questions is a big question. To answer a big question you need a big mirror. A big telescope mirror will help you unlock the secrets of the universe. But we don't have many big telescopes. The combined collecting area of the mirrors of all the big telescopes in the world comes to roughly the size of a doubles tennis court. But by the turn of the century, we'll have new telescopes that will give us about four times this area.

The astonomers call a telescope a 'light bucket'—something that collects light. A telescope is like a camera stuck onto the end of a telephoto lens. But the big telescopes don't use lenses. Once you get a lens bigger than a metre across, it sags in the middle because of its own weight. And you can't prop a lens up in the middle, otherwise you block the light. But mirrors can be held up from the back, not just the sides, and that's why they use mirrors instead of lenses in all the big telescopes.

One of the biggest telescopes around is the Mt Palomar Telescope in California. Its mirror is about 5 metres across, and took about 11 years to grind to the right shape.

The Palomar saw first light way back in 1947.

Since then, most of the new large telescopes that have been built have been about 4 metres across. There's been no real point in building bigger telescopes, because the astronomers have been getting more efficient at using the light which falls into their light buckets. Back in 1948, the best photographic film available picked up less than 1 per cent of the incoming light. But now we don't use film much anymore. Instead we use electronic light detectors like charge-coupled devices (CCDs). These CCDs can pick up about 70 per cent of the incoming light. That's close to the theoretical limit, so it's now time to build bigger telescopes, if we want bigger answers.

But it's easy to make mistakes when you build a big telescope. The Soviets built the world's largest telescope. It's in the Causasus Mountains, and has a mirror 6 metres across. This telescope has never given high quality images—but it stayed on the ground and got fixed (unlike the Hubble Space Telescope)! For one thing the mirror was not ground to a very smooth shape, it was too bumpy. Another problem is that the mirror is solid, not a hollow honeycomb structure like the Palomar 5-metre mirror. This mirror is so heavy, that it's like a thick solid brick wall. It takes all night to get rid of the heat that it picked up during the day. A mirror only gives a good sharp picture when the whole mirror is at the same temperature—otherwise there are shimmering heat waves. The final problem with the unfortunate Soviet 6-metre telescope, is that the tempestuous wind currents around the Caucasus Mountains blur the image.

A different approach was taken with the new American Hubble Space Telescope. This telescope has cost about $1.2 billion, about 30 times the cost of an observatory down here on the ground, and it has taken roughly half a working life time to get up and running. It's in a low earth orbit around our planet. Its images aren't affected by the moving atmosphere, and so they should have been much sharper than any ground-based telescope.

But there was a terrible mistake. An astronomical telescope has two mirrors—a big one that catches the light from space, and a smaller mirror that catches the light from the big mirror and then sends it to the CCD. The 2.4-metre primary mirror is the smoothest mirror ever made in the history of the human race, and the secondary mirror is almost as good. But the two mirrors are not properly matched to each other—and so the photos will not be as sharp as they could have been.

The makers didn't ever test the pair of mirrors on the ground, because it would cost too much—several hundreds of millions of dollars. But the military get the money they need to test their KH-11 spy telescopes that can pick up a golf ball on the ground—and a KH-11 is just a Hubble that is designed for close work, and that is pointed at the ground, not at the stars.

Quite a different approach again was taken with the Keck Observatory, 4 kilometres above the sea level on a dead volcano in Hawaii. This observatory is on the highest point in the Pacific Ocean, and the air is so thin that the observers have to acclimatise. There are even bottles of pure oxygen to use in emergencies. This telescope has a mirror 10 metres across, which is roughly the width of a tennis court. It cost $87 million which is a lot less than the $1.2 billion of the Hubble Space Telescope.

This 10-metre mirror will have three times the light-collecting area of the 6-metre Soviet telescope. A revolutionary design had to be used for this telescope. A single one-piece mirror 10 metres across would be very thick and heavy, and would tend to sag when it was tilted away from the vertical. It would also have needed very large, heavy and expensive bearings on which to mount the telescope. So the designers went for a multiple mirror design.

This telescope has 36 separate hexagonal mirrors, each of them 1.8 metres across. Each mirror is about 75 millimetres thick, so the glass will weigh a total of 14 tonnes. That's roughly the weight of the Palomar 5-metre mirror, which is about four times smaller.

The mirrors are arranged into three concentric rings. There are six mirrors in the innermost ring, 12 mirrors in the second ring around that and 18 mirrors in the outside ring. Each glass mirror had to be individually ground to exactly the right shape.

And just to make sure that they keep that right shape, as the telescope tilts away from the vertical, computer-controlled 'active optics' continually check the performance of each mirror and its shape is then individually adjusted by motors and levers on its back. This is brand-new 'Star Wars' technology. It means that all the mirrors together make one continuous surface, 10 metres across, but smooth to 50 billionths of a metre, or a thousandth of the thickness of a sheet of paper.

There's even a weird new telescope with a liquid mirror. It's being made by Ermanno F. Borra of Laval University in Quebec. The mirror is made out of a few thousand dollars worth of liquid mercury. Now the perfect shape for a telescope mirror is a parabola. The mirror inside your car headlights has a shape very close to a parabola. The beauty of a mirror in the shape of a parabola, is that if you focus on a very distant star, the parabola shape brings all the rays of light to a single focus point, so you get a very sharp image. And by a very nice coincidence of physics, the surface of a liquid in a spinning bowl has the shape of a parabola. You can prove this for yourself by putting a bowl of water on your record turntable, and running it at the lowest speed. The centrifugal forces push the water towards the edges of a bowl, and into that perfect parabola shape.

So Ermanno Borra made a fibreglass mould in the rough shape of a parabola, and put a small pool of mercury in the bottom. He spun the fibreglass, and the mercury dipped down in the centre, and climbed up the sides of the bowl to take on that perfect parabola shape. The advantage of this mercury mirror is that it is very cheap. But the big disadvantage is that you get this perfect parabola shape only when the mirror faces straight up. Otherwise, gravity makes the liquid mirror sag.

It is not a new idea. Spinning mercury telescopes were made in Europe in the middle 1800s. But Robert W. Wood, a professor of physics at John Hopkins University in 1909 was one of the most successful. He built a half-metre mercury mirror. But he couldn't isolate his spinning mercury telescope from vibrations, so the 'Footsteps of a person running 50 yards from the telescope house'

would make the mercury ripple.

But Borra had the advantage of modern mechanical technology. He built a 1-metre mercury mirror in 1982. To keep the surface of the mercury free of ripples, he used a super-low friction air bearing and the same sort of belt drive to run his turntable that you find in many high quality record turntables. He also used a synchronous motor driven by the frequency of the driving voltage. And to finish it off, the actual mercury is covered with a thin layer of transparent mylar, to protect the mercury from air drafts. The mercury was about 3.5 millimetres thick while the mirror was spinning.

Since then Borra has built a 1.5-metre mirror which spins about one revolution every four to five seconds. It costs about $15 000, much less than $100 000 for a conventional glass mirror.

But the most revolutionary approach was taken by Roger Angel of the University of Arizona. Roger loves simplicity. He reckons that a large mirror should be made out of a single lump of glass, not a whole bunch of small segments. He also reckons that a large mirror should be strong enough to support itself without having to use fancy 'active optics'. And he also reckons that you shouldn't waste years grinding glass.

So he uses two tricks. First, he uses a honeycomb design in the back of his hollow mirror to save weight, but still keep the mirror stiff. Second, he uses a spinning turntable inside the furnace, so that when the mirror comes out of the furnace, it's close to the final shape. For a 3.5-metre mirror, this means that instead of having to slowly grind out 10 tonnes of glass, they have to to remove only 400 kilograms. And his hollow 3.5-metre mirror weighs only 2 tonnes, as compared to 11 tonnes for a solid mirror. The bees were on to a good thing when they invented the honeycomb structure for their

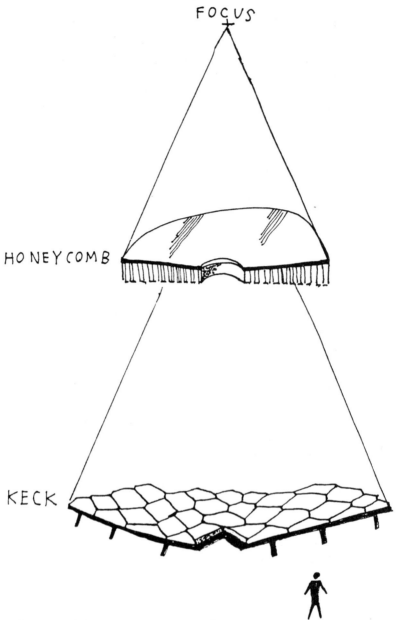

bee hives to make them light and strong.

But Roger Angel gets really revolutionary when he casts his lightweight mirrors in a revolving furnace. His rotating furnace weighs about 120 tonnes, and inside the furnace is a turntable about 12 metres across, roughly the size of a small house. And sitting on top of the turntable is the mould.

To make a 3.5-metre mirror they start off by loading individual 1-kilogram chunks of glass. It takes more than three days to load in the 2.4 tonnes of glass, because each chunk has to be individually checked for flaws. They then close the lid and feed in 200 kilowatts of electricity to heat up the furnace. At around 900°C the glass begins to run into the mould, and the turntable starts rotating at 8.5 rpm. The temperature climbs to 1200°C, hot enough to thoroughly melt the glass, and is held at this temperature for three hours. Large trapped bubbles of air rise to the surface and pop. The furnace needs only 130 kilowatts to keep it hot.

Because the turntable with the molten glass is spinning, the glass takes up the shape of a parabola. The heat is then turned off, and the furnace keeps on spinning for about 24 hours, until the glass cools to about 700°C. At this temperature, the perfect parabolic shape is frozen into the glass. The lid of the furnace is lifted and the mirror is inspected. If all is OK, the lid is closed, and the mirror is cooled down at the very slow rate of less than 0.5°C per hour. If it cools any faster, the mirror could crack. The whole process takes about six weeks.

The mirror is already pretty close to the perfect parabolic shape. Because the glass comes out of the furnace with a dished shape, not a flat shape, they can reduce the time for grinding from many years to a few months. If you began with a flat disc of glass, instead of one that came precurved out of the furnace, you would have to spend years grinding away 30 tonnes of glass to end up with a 14-tonne 8-metre mirror.

Once the glass has been ground to the right shape, it's coated with a thin layer of reflective aluminium, and then it's ready to go to work as a mirror, unlocking the secrets of the universe.

Angel's revolutionary new technique of casting lightweight honeycomb glass blanks on a spinning turntable inside a furnace has already paid off. In his first stage, he made a 1.2-metre mirror for the Smithsonian Institution, and a 1.8-metre mirror for the Vatican. It's nice that an Angel made a mirror for the Vatican to explore the heavens.

In his second stage, he made three mirrors each 3.5 metres across. Two of these mirrors are going into telescopes, while the third is being used by the United States Air Force to test Star Wars lasers. The next stage will be a 6.5-metre mirror to replace the six separate 1.8-metre mirrors on the Multiple Mirror Telescope on Mt Hopkins near Tucson, Arizona. The changeover will cost about $11 million, and should be finished by mid-1993.

If there are no problems, the final stage will be to make the enormous 8-metre mirrors, for which he'll have to build a new furnace. He wants to turn out these 8-metre mirrors at a bargain-basement cost of just $5 million.

The first two 8-metre mirrors will be installed into the world's largest pair of binoculars on Mt Graham in Arizona. The light from the two telescopes will be brought to a single focus, giving this giant set of binoculars the light-gathering ability of a single 11-metre telescope. The cost will be only about $50 million. The third mirror has already been earmarked to go into a new observatory high in the Chilean Andes, where the air is clean.

Once Angel's lightweight mirrors hit the market, eveyone will start to make them. Even an amateur will be able to get a cheap 1.5-metre mirror. With so many people looking into the heavens, we could get the answers to questions that have not even yet been asked.

REFERENCES

Popular Science October 1987 'Spinning Scopes' by Arthur Fisher, pp 76–79, 101, 102
New Scientist No 1636, 9 October 1988 'The Great Telescope Race' by Nigel Henbest, pp 52–59
Astronomy July 1988 'The Telescope that Defies Gravity' by Richard Berry, pp 42–47
Astronomy Now April 1989 'Spin-Casting Magic Spells Big Mirrors' by Simon Mitton, pp 26–31
Discover July 1989 'The Big Glass' by Terry Dunkle, pp 69–81

Pluto

*P*luto is the most mysterious planet in our solar system. It was not discovered by a professional astronomer, but by a farm boy in 1930. Back then, Pluto was in the right place, but for the wrong reasons.

The early astronomers and astrologers knew about the first six planets long before the human race invented writing. Those planets were Mercury, Venus, Earth, Mars, Jupiter and Saturn. In 1781 William Herschel used a telescope to discover a new planet past Saturn, the seventh planet which we now call Uranus. In Greek mythology Uranus was the father of Saturn.

But over the next half century or so, astronomers saw that Uranus was not travelling at a smooth speed through its orbit. They thought that it was being pulled by a heavy unseen eighth planet. Two mathematicians, John Couch Adams in England and Urbain Leverrier in France independently sharpened their pencils and worked out just where this planet should be. In 1846, these mathematical results were sent to Johann Galle, an astronomer at the Berlin Observatory and within two days he found the eighth planet within a few degrees of its predicted position. He called it Neptune, after the god of the sea, because of its beautiful blue-green colour.

Within another half century, astronomers were up against the same unnatural behaviour again. Not only was Uranus still wandering in a bumpy orbit, so was Neptune. It was time to send in the mathematicians.

Percival Lowell was an eccentric Boston citizen who was mad about astronomy. He was wealthy enough to build and maintain a professional observatory in Flagstaff, Arizona. He aimed its telescopes at Mars, the fourth planet, and found canals—and we all know about the non-existent canals of Mars.

But he also looked for the ninth planet, both with pencil and paper, and his new telescopes. He guessed that the new planet had about six times as much mass as the Earth, and so using this as a first assumption, he sharpened his pencils and came up with a location.

His first unsuccessful search for this mysterious ninth planet ran from 1905 to 1909. It was a slow process. Each night he would expose a photograph of a small piece of sky, near where his calculations said the planet would be. Each exposure took a few hours because the films that were available then were very slow.

Then a few days later he rephotographed exactly the same piece of sky. He would stack the two negatives on top of each other, and look for any differences between them with a magnifying glass. Anything that had changed position could be his mysterious ninth planet—unless it was an asteroid, or a meteor. This technique was too slow, so he got a machine called a blink comparator.

A blink comparator switches vision very rapidly between two photographs. Any object that has changed position over the few days when the photographs were taken appears to blink out of existence in one position, and blink into existence in another position. Now that Lowell could search photos rapidly, he began to work in earnest. He hired several people and formed a team, and they began to work on the long and boring calculations needed to work out a more accurate position for the new planet. Mind you, a modern hand-held calculator could do in a few minutes what it took the team many months to do.

His next survey of the sky began in March 1911. As he continued to fine tune his predictions, he kept on sending his results to his observatory in Arizona. Pluto was actually photographed in 1915 on two occasions— 19 March and 7 April. But whoever was checking the photos must have been half asleep and didn't notice the ninth planet. They missed out on their claim to fame.

In 1916, Percival Lowell died. His will left $1 million to the observatory, but his widow, Constance Lowell, contested the will and delayed the search for ten years.

By 1929, the observatory had installed a new and larger telescope, but it had also run out of money. Luckily, Clyde Tombaugh, a 22-year-old Kansas farm boy, applied to work at Lowell's observatory in Arizona. In his application he sent drawings of various objects that he had seen in the heavens with his own homemade telescope. He was a stroke of luck. He had a good sharp eye, he could keep the fires burning and clean out the offices, and best of all, because he was not a professional astronomer they wouldn't have to pay him much at all. The observatory hired

him immediately.

He settled into the new routine rapidly. At night he would photograph the skies, and by day he would check the photos for movement. Each plate had half a million stars. On 1 April 1929, Pluto appeared on one of his plates, but he did not notice it. But fate gave him a second chance and on 18 February 1930, he found a tiny dot that moved the right amount. He had done what had been done only twice before in recorded human history—he had discovered a new planet. He told his boss, and that night went to see Gary Cooper starring in a new movie called *The Virginian*.

But what to call the new planet? Percival Lowell's widow Constance suggested four names—Zeus, Percival, Lowell and finally her own name, Constance. But the Astronomical Union ignored her suggestions, and called this planet Pluto after the dark god of the underworld, the world of eternal darkness. It's a good name, because firstly it's very dark in the outer solar system, and secondly the first two letters of Pluto were Percival Lowell's initials.

Pluto was only a few degrees away from the predicted location, but there was a small problem. Pluto was much too small and light to have caused the bumps in the orbits of those giant planets Uranus and Neptune— but by an amazing coincidence, Clyde found it in the predicted place.

The next major discovery about Pluto was made in 1979, when James Christy of the United States Naval Observatory suddenly noticed that photographs of Pluto had a bulge on one side. And when he had a look at photographs taken a few days before, he noticed the bulge was on the other side. Pluto had a moon! It spun around Pluto every 6.4 days.

He named this moon after Charon (pronounced karon), the ferryman in Greek mythology who rowed souls across the river Styx from the world of the living to Pluto's world of the dead, Hades. But because his wife, Charlene was known as Char, he proposed that in future everyone should pronounce karon as shar-on. His delighted wife said 'Some husbands promise their wives the moon, but mine got it for me'.

The next great series of discoveries began in 1985. From 1985 to October 1990, Pluto and Charon were lined up with the Earth. This means that they eclipsed each other every 3.2 days. This was a fantastic opportunity, and would not happen for another 124 years. So the astronomers would see both Pluto and Charon, but also they could see Pluto in front of Charon or vice versa.

They discovered that Pluto was about 2300 kilometres across and nearly twice as dense as water. About 75 per cent of Pluto is rock, about 20 per cent is water-ice, and about 5 per cent is methane-ice. Pluto probably had some sort of meltdown in its early days, when it got very hot. One-third of the ice turned to water and most of the rock sank to the centre. So Pluto has a rocky core, with a layer of water-ice and with methane on the top.

Pluto is basically a dark reddish colour around the equator, which is at a temperature of about 59 degrees above absolute zero (−214°C). This red colour is due to old methane-ice probably mixed in with a few organic chemicals. The polar caps are about four times brighter and quite light in colour because of the new fresh methane-ice. This ice extends from the poles down half way to Pluto's equator, and is at a temperature of about 54 degrees above absolute zero, (−219°C). Pluto reflects about 60 per cent of the light that lands on it, which is why Tombaugh was able to photograph it.

Charon is about half the diameter of Pluto, about 1300 kilometres across. It is greyish and seems to have absolutely no methane on it, only water-ice. And at what would be the temperate zones down here on Earth, there are two bands—one band is dark and the other is light. Charon reflects 40 per cent of the light that lands on it. Because Charon is about half the diameter of Pluto, we shouldn't call them a planet and moon, but a double-planet system.

The next major discovery for Pluto happened on 9 June 1988. For a period of only 18 seconds, just after midnight, Pluto passed in front of a very faint star. This eclipse was visible from the South Pacific region, including Australia and New Zealand. Temporary astronomical observatories were set up in Australia and New Zealand. NASA called its flying observatory, a converted Lockheed transport plane, into action as well.

The astronomers noticed that as Pluto swung in front of this distant star, the light didn't switch off suddenly, but dimmed over a few seconds. And then the light brightened again slowly on the other side. They reckon that this means that Pluto has an atmosphere. The atmosphere is in two layers; it has a layer 50 kilometres thick close to the surface, and a thinner layer 300 kilometres thick on top of that. But the 'air' is a few million times thinner than the Earth's atmosphere. The temperature of the atmosphere is about 68 degrees above absolute zero (−205°C).

Pluto takes around 248 years to travel around the sun, but for about 20 years of each orbit, it cuts inside the orbit of Neptune. It last did this in 1979, and had its closest approach to the sun in 1989. It is now heading out and will cross the orbit of Neptune in 1999 to become again the most distant planet in the solar system.

But Pluto will never collide with Neptune. Pluto has a very tilted orbit, so it always misses Neptune. And the two planets are locked together into a stable resonance orbit. While Pluto takes 248 years to go around the sun, Neptune takes exactly two-thirds as long. So every time Neptune does three revo-

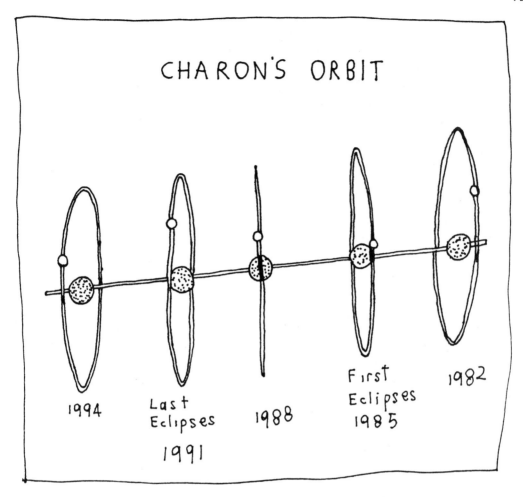

CHARON'S ORBIT

1994   Last Eclipses   1988   First Eclipses 1985   1982

1991

lutions of the sun, Pluto has done two orbits, and they're back to their starting positions relative to each other.

But we are still left with the problem of Pluto being too light. Pluto was discovered because the astronomers thought that a planet roughly six times more massive than the Earth was pulling Uranus and Neptune slightly out of their orbits. But Pluto is not six times more massive than the Earth, it's actually 1/500th as massive. The strange thing is that Pluto was in almost exactly the right position, even though it was the wrong mass. This probably happened by chance. Many astronomers think that there is still another body out there, gently tugging at Uranus and Neptune with its gravity. Pluto has finally come in from the cold.

REFERENCES

Science Vol 237, 31 July 1987 'Improved Orbital and Physical Parameters for Pluto–Charon System' by David J. Thalen, Mark W. Buie, Richard P. Binzel, and Marion L. Frueh, pp 512–514
Astronomy September 1987, pp 76, 77
Science Vol 237, 11 September 1987 'The Surface Composition of Charon; Tentative Identification of Water-Ice' by Robert L. Marcialias, George H. Rickie, Larry A. Lebofsky, pp 1349–1351
Science Vol 237, 11 September 1987 'IRAS Serendipitious Observations of Pluto and Charon' by Mark V. Sykes, Rock M. Cutri, Larry A. Lebofsky, and Richard P. Binzel, pp 1336–1340
Astronomy September 1988, pp 52, 53
New Scientist No 1662, 29 April 1989, pp 21–26
Technology Review July 1989, pp 9–10

Potatoes can Kill

*P*otatoes can kill humans, especially the unborn. And thanks to the potato, the Irish are scattered across the world.

Potatoes have been cultivated for about 8000 years. The Peruvians have been freeze-drying potatoes for over 2000 years. But 450 years ago, the Spanish conquistadors came tramping across the ground looking for the fabulous gold and silver of the Incas and they ignored the humble potatoes in the dirt. The Inca empire was crushed, the Spanish empire has faded away, but King Potato still reigns supreme.

Each year 300 million tonnes of potatoes are harvested, with a total value of about $200 billion. That is more than the value of all the gold and silver that the Spaniards ever stole from the Incas.

But the Spaniards ran into great resistance when they tried to introduce the potato into Europe. The Scots would not touch the potato, because it was not mentioned in the Bible. But eventually the potato was accepted and became essential. In fact, the Industrial Revolution could not have happened without the potato. It was a cheap and hardy source of food for the new workers funnelled into the cities.

But in 1845 a disaster hit the potato. In Ireland 1845 was a wet and humid year. A fungus grew over the potatoes, which soon began to die. Out of 5000 different types of potatoes available, only two types had been brought to Ireland. These two types were both susceptible to the fungus. One million out of the 12 million people living in Ireland died from starvation. Another million emigrated out of Ireland, looking for a decent meal. So we can thank the potato for all the Irish people in Australia and America.

With potatoes, you can grow more nutrients on worse land, and harvest it faster than you can with any other crop. And potatoes will grow from below sea level in the Netherlands, to over 4000 metres above sea level. A potato is very nutritious but one-third of all the nutrients are immediately under the skin. So your mother was right— 'eat the skin, it's the best part'. There was a Scandinavian man who lived for 10 months on nothing but potatoes, with just the tiniest smear of margarine. And to graduate from some French cooking schools, you have to be able to make 60 different potato dishes. The potato is 99.9 per cent fat free—it is the stuff that you put on it that is fattening. A single spud has more than half of the vitamin C that

PRESS RELEASE:

KING SPUD VERSUS THE WORLD!!!

INTERNATIONAL PERSONALITY KING SPUD HAS FINALLY
BEEN EXPOSED:- DISGUISED AS A POTATO HE SECRETLY
EXPOSES HIMSELF TO SUNLIGHT AND WITH ONLY
ONE BITE DISPOSES OF THE ENEMY BY TURNING
GREEN . . . .
                    NA NA NANA- NA NA NANA . . . .

BE WARE

you need each day. Of course the word spud comes from the name of the small spade used to dig it out with.

Everybody hates to peel potatoes, so the fast-food industry had to come up with an industrial-strength method of getting the skin off—and it's not very pretty. When they make chips, as in fish-and-chips, they explode off the skin. The potato is put into a giant pressure cooker and boiled at a pressure of 15 atmospheres. Any moisture under the skin of the potato would love to expand and turn into steam, but it's stopped by this external pressure. But suddenly, the external pressure is released, the water under the skin expands, and the skin explodes off the potato like a hand grenade.

And how do they turn these naked shell-shocked potatoes into chips? They put them into a pipe with water in it flowing at 80 kilometres per hour. The potato then commits hari-kiri by plunging itself onto a set of knife blades, turning itself into a bunch of skinny chips.

But potatoes can be poisonous, especially when they are green. The chemical is called solanine, and all potatoes have a tiny bit of it. Solanine is as poisonous as strychnine. At least 30 people have been killed, and 2000 suffered food poisoning from this solanine.

Now if you leave the potato in the sunlight for about two days, it will turn green. This green colour is due to the chlorophyll. The chlorophyll is not poisonous, but is made at the same rate as the solanine. So if a potato

has more green colour, it has more solanine. And if you are rough with your potatoes and drop them and bruise them, in defence, they produce more solanine as well. But it's solanine that gives potatoes their taste. The chemical does not build up in your body; it just washes out each day. But if you have too much solanine in one dose, it will cause vomiting and diarrhoea, and occasionally death. If you want to try the taste of pure solanine, try having a little tiny nip of the end of a green shoot on a potato. That green shoot has as much solanine in it as an entire spud.

But don't try it if you're pregnant. Solanine is very dangerous for pregnant women. The foetus can die while it is still inside the uterus, leading to a miscarriage. And sometimes the foetus just slowly vanishes and fades away, reabsorbed by the uterus.

So avoid potatoes that are green or have shoots on them—do not try and cut away the green bits, just toss away the whole potato. And be especially careful if you are a woman who is pregnant or likely to become pregnant. If your greengrocer is selling green potatoes, don't be green enough to buy them.

REFERENCES

National Geographic May 1982, pp 668–694
Food Technology in Australia Vol 36 (3) March 1984 'The Toxicity and Teratogenicity of Solanaceae glycoalkalids particularly those of the Potato (*Solanus Tuberosus*)' a review by C. Morris and T. H. Lee, pp 118–124
Chemistry in the Market Place by Ben Selinger, 3rd edition, Harcourt Brace Jovanovich, Australia 1986, pp 70, 367–368, 371
New Scientist No 1615, 2 June 1988, pp 54–57
Science Digest December 1988, pp 26, 109

Reading makes you Short-Sighted

*S*ince white man taught the Eskimos to read, they have had to start wearing glasses.

In fact, scientific research, backed by statistics and short-sighted chooks, shows that reading can be bad for you, and even make your eyes grow into funny shapes. On the other hand, if you are a male and are short-sighted, you tend to be more intelligent, and have more years of schooling.

If you have average vision, you can see about two million light years. That's how far away the galaxy Andromeda is, and it's the furthest object that you can see with the naked eye. If you want to see further, you have to use a telescope. But if you're short-sighted, Andromeda would be so fuzzy that it would just fade into the background. Short-sight means that you can see things clearly only at a short range, and that your long vision is blurry. Long-sighted people can see Andromeda clearly, but they have trouble reading books with the naked eye.

The human eye is a sphere roughly 25 millimetres in diameter. It's always round because a pump keeps the fluid inside the eye at a constant pressure. The retina is a layer wrapped around inside the sphere, and it's about one-third of a millimetre thick, or about four times the thickness of a human hair. The retina is like the film in a camera— it takes the pictures for the brain. The retina actually senses the light and turns it into electrical signals. The brain then turns these electrical signals into what it foolishly thinks is reality.

If you look at somebody's eye from the side, you'll see a jelly-like bulge. This is called the cornea and it helps bend the incoming light rays so that they form a clear image on the retina. If you want to have perfect vision, the image has to land exactly on top of the retina. And if it doesn't, what you see is fuzzy.

When we're born, our eyes are too small. The head is a tight fit through the birth canal, and every little bit helps. Babies are all a little long-sighted, and the images of the outside world tend to land behind the retina. For a youngie, close-ups like faces and rattles are very blurry. As we grow older the eye gets bigger and reality gradually gets sharper. But if the eye kept on growing reality would get

blurry again. What tells the eye when to stop growing?

Josh Wallman, a biologist at the City University of New York, did some experiments with baby chickens. They also start life with small eyes. In his experiments he covered one eye of newborn chickens with miniature ping-pong balls. So with the covered eye, the chicks could see the difference between light and dark, but they could not see pictures or shapes. He left the other eye alone. After six weeks he measured the size of both eyes. The uncovered eye had stopped growing at just the right size, and had perfect vision. The ping-pong ball eye had kept on growing and it had become so short-sighted, it would never see without glasses.

It seemed to Josh, that if an eye did not get clear pictures on the retina, it would keep on growing, in a hopeless attempt to bring the fuzzy reality into focus. Just as the actress said to the bishop 'If you don't use it, you lose it'.

But suppose that the image of the outside world lands exactly on the retina and we get a clear picture of reality. Now the retina is really getting stimulated from the hard edges. The theory is that the retina then puts out some sort of chemical that says, 'slow down, no more growing, I'm perfect, this eyeball is exactly the right size'.

To test the theory a bit more, Josh Wallman did another experiment with some more cute baby chickens. This time he covered only part of one eye. He found that the eyes grew strange bulges, but only in those parts where they were covered up.

So this is Josh's theory, based on the fact that babies' eyes start off small like chickens' eyes, to explain why reading makes you short-sighted. When you read, you see only one word at a time in sharp focus, and the rest of the page is a blur. Now if you spend a lot of time reading as a young person, only a very tiny bit in the middle of the back of your eye is stimulated. The rest of the retina just sees an out-of-focus blur, so it keeps on growing. The much-larger growing section of the eyeball drags the tiny central section along with it. So your eyeball grows too big, and you become short-sighted—condemned to never seeing Andromeda clearly with the naked eye. But luckily, in the 1500s people began to wear bits of transparent rock on their faces—and so with glasses, almost anybody can see the stars.

It might be a long conclusion to draw from an experiment with a few half-short-sighted baby chickens, but it does make sense that a lot of reading can make you short-sighted. The Chinese have had a cure for this for thousands of years—every hour or so, children at school who are reading have to stop and spend five or so minutes concentrating on far away objects.

So, as the far-sighted Chinese visionaries would say—there's more than one thing that you can hold in the palm of your hand that can make you go blind.

REFERENCES

Science 3 July 1987, Vol 237, pp 73–77
Scientific American October 1987, p 23
Omni December 1987, p 49
Popular Science February 1988, p 10
Nature Vol 333, 23 June 1988 'Eye Development and Short Sight' p 707

Rogue Waves

*O*n 3 June 1984, 39 tall ships were taking part in a race across the Atlantic Ocean. Just before sunrise, the 36-metre, three-masted square rigger *Marques* was about 120 kilometres north of Bermuda. Fierce rain and an angry wind sprang up. Most of the crew of 28 were asleep below deck.

The *Marques* was built to survive conditions worse than this, but the captain shortened the sails, just to be safe. Suddenly, a gust of wind pushed the *Marques* right over to one side. While the ship was down and before it had a chance to recover, a gigantic wave of unbelievable force and size slammed on top of the already tilted ship, and shoved its masts underwater. Seconds later, another wave slammed into the ship, and delivered the killer blow. The *Marques* vanished under the waves in less than 60 seconds—and only nine of the 28 crew survived.

The ship should have recovered from the sudden gust of wind that pushed it over, but those two huge and unexpected killer waves delivered the final blow. Because of their reckless and unpredictable behaviour, sailors call them rogue waves.

An unexpected huge rogue wave flipped a 12.5-metre trimaran that left the South Island of New Zealand. The boat was on a three-week voyage to Tonga, but the rogue wave turned the trimaran upside-down on the fourth day out. The four crewmen were trapped in the cabin, but managed to smash a hole in the hull and climb out. They lived on the hull of the boat for 119 days, catching and eating fish, and collecting rainwater with sheets of plastic. They had lost a lot of weight, but were basically well and healthy when the still-inverted boat smashed up on Great Barrier Island, about 700 kilometres away from their original encounter with the rogue wave.

We have had legends about rogue waves ever since humans cast out to sea. Two things make them very dangerous—their huge size, and the fact that they're totally unpredictable. Over the last 10 years 460 ships have simply vanished at sea. They were not small yachts—they all displaced more than 500 tonnes.

Rogue waves can pop up out of a flat sea—they're not a series of large waves, they're just one or two solitary crests. They can turn a ship over, or worse still, snap a ship in half like a carrot. But scientists are finally beginning to understand how they happen.

Waves come mostly from the wind, with a

DON'T WORRY WE'LL MAKE NEW YORK BY THE MORNING!

bit of help from the tides, storms, and currents. Out in the ocean, waves behave very differently than on the beach. They travel in different directions and at different speeds, and they have different sizes. Usually, the waves are out of step with each other, unlike the breakers at Malibu.

The old theory to explain rogue waves was all about waves getting in step. Sometimes one wave will be a mirror image or upside-down copy of another wave that it collides with—so the peaks of one wave will match up with the troughs of the other wave. When this happens, the waves cancel out, and a short-lived patch of flattish water appears.

But sometimes the waves get in step with each other, like marching soldiers on a bridge. When the peak of one wave lines up with the peak of another wave, they make a new wave that is as high as the two smaller waves added together.

At the National Research Council of Canada's Hydraulics Laboratory in Ottawa, they have a huge indoor tank, the size of an Olympic pool, with 60 separate computer-controlled paddles run by some splashing software. These paddles can make absolutely any sort of wave. They've found that the old theory was right—that when waves get in step with each other, they can make huge rogue waves. But they've also found that rogue waves are more likely to occur in certain locations.

For example, there has long been the myth that in high seas it is safe to hide behind the shelter of an island. But this can sometimes be a fatal mistake. When waves approach an island, they split up to go around the island, and then meet on the other side. And when they do meet up, depending on what the local sea floor looks like, instead of sailing into an area of relative calm, you can sometimes find yourself in a zone of unexpected lethal rogue waves.

The highest rogue wave ever reported was measured by the crew of the US Navy ship *Ramapo* in 1933 in the Pacific Ocean. It measured 35 metres from the trough to the crest, taller than a 10-storey building. In World War II the *Queen Elizabeth* was severely jostled by a huge rogue wave when she was working as a troopship near Greenland. The enormous ship was so badly shaken, that three people died from injuries. In 1974, a rogue wave off the coast of South Africa actually tore off the entire bow of a Norwegian tanker.

Small ships can be completely overwhelmed by a rogue wave, unless they're lucky enough to ride over it, like Captain Goodvibes catching a monster wave. But large ships can simply crack into pieces if the bow and stern are left dangling in the air, while the crest of a rogue wave lifts up the ship in the middle.

Now there's only one safe place from which to look for rogue waves, and that's the sky. The first satellite to monitor the oceans, Seasat, was launched for the US Navy in 1978. The oceanographers gleaned more knowledge about the currents, eddies and waves in the oceans in the first three months of Seasat than ever before. But there was a rumour that Seasat was so sensitive that it could even pick up the trails of nuclear submarines while they were under the waves. Suddenly, Seasat broke down, and mysteriously, was never replaced.

But now commercial non-military satellites will be launched over the next five years to monitor the surface of the sea. After all, if the ill-fated *Marques* had been only 150 metres away from where it was hit, it would probably have survived. Once we know where, when and how these rogue waves occur, ships will be able to sidestep this scoundrel of the sea.

REFERENCES

Discover April 1989, pp 47–52
Van Nostrand's Scientific Encyclopaedia pp 2060–2064
Time 16 October 1989, p 29

Sex of your Children

*e*ach year three out of every 200 Australians will become either a mother or a father.

Eighty per cent of those couples will choose to marry. Until recently they couldn't choose the sex of their new child. But now you can, and sometimes the choice has already been made for you—by your job, or the amount of hair on your head.

More boys are born than girls. In every 1000 newborn Australian children there are 30 more boys. But the male of the species is weaker, and more boys die in that first year of life. It's nature's way of compensating, so the world ends up with a roughly equal mix.

At least it used to, back in the 1930s. Then, beef farmers didn't have to feed hormones to their cattle to fatten them up quickly, and butchers' wives were giving birth to kids in the same ratio as everybody else—106 boys for every 100 girls.

But in the 1970s there were piles of tender meat rich in female hormones on display in the shops. Butchers were soaking up female sex hormones straight through their skin as they carved the cuts. They only found out years later that their wives were having more girls than boys thanks to the female hormones. To tip the balance further, butchers eat more meat than the rest of us. Some of them even grew breasts.

In the 1980s, the era of greed, farmers found that new artificial male hormones made cattle grow even faster and fatter again. So the new occupational hazard for butchers became male sex hormones. Today American butchers' wives are having 121 boys for every 100 girls. That's 95 extra males in every 1000 newborn kids.

But it's not just the butchers who are missing out on equal rights. Blokes who either drink a lot or work in the alcohol industry tend to father girls. This includes virtually everyone from cellar workers to waiters. It seems that these workers don't mind the occasional drop of the medicinal, and alcohol reduces your production of the male sex hormone, testosterone. Any imbalance in testosterone causes girl babies.

Jet fighter pilots like Tom Cruise have more daughters than sons. In fact, 60 per cent of 'top guns' have girls. Flying fast makes females, and it's probably the high G-forces in the cockpit that do it. G-forces are like an artificial gravity. They can force all the blood down to the feet, so that the pilot blacks out. If the blood is forced to the head, the pilot has a red-out. The pilots have to wear special G-suits. G-suits have inflatable rubber bags in the legs, trunk and arms that automatically inflate in a tight turn to push the blood

back to where it came from. During some manoeuvres, a fighter pilot has to endure up to nine times the weight of gravity on everything in his body, including the tiny sperm cells. So maybe G-force should really stand for girl-force, not gravitational-force. And that movie about pilots (the men with 'The Right Stuff' which was called *The Wild Blue Yonder* should really have been called *The Wild Pink Yonder.*

But it's not permanent. Bert Little, a geneticist in Dallas, Texas, found that if pilots stop flying for 90 days, then they produce more boys again. It takes the testicles about 72 days to make a new batch of sperm. One explanation of the girl-force is based on the fact that not all sperm are equal. Boy-making wrigglers are slightly smaller than girl-making wrigglers and might be weaker and die off under the stress.

But on the other hand, Japanese scientists have used this weight difference. They can pick the sex of your next child, and it's as easy as separating oil from water. They put the sperm in a high-speed centrifuge, which makes a sort of artificial gravitational field. They end up with two layers in the centrifuge, with the heavier X girl-making sperm on the outside. The parents then choose the layer, and can go shopping knowing the colour of the baby's clothes.

But you don't always need a centrifuge, because the amount of hair on your head can pick the sex of your children. If a man has a very high level of the male sex hormone, testosterone, it's likely that he will go bald early in life, and it's also likely that he will father daughters.

Not all creatures have their sex worked out as soon as egg meets sperm. The sex of the baby reptiles depends on the temperature that the egg is exposed to after it is buried in the ground. So if mummy crocodile or mummy lizard buries the eggs close to the warm surface, the babies will probably be boys, and if she buries them cool and deep, they'll hatch out as girls. It's the other way around for turtles. So in these reptiles, it's not the heat of the moment, it's the heat of the after-moment. If we were reptiles instead of mammals, we could work out the sex of our future children with a thermometer.

REFERENCES

Australian Doctor 3 April 1987, p 15
Aviation, Space and Environmental Medicine Vol 58, July 1987 'Pilot and Astronaut Offspring: Possible G-Force Effects on Human Sex Ratio' by Bertis B. Little, Cecil H. Rigsby and Lori R. Little, pp 707–709
Nature Vol 329, 17 September 1987 'Environmental Determination of Sex in the Reptiles' pp 198–199
New Scientist No 1657, 25 March 1989, pp 33–38

The Shark with the Fire in its Belly

*t*he great white shark is the world's largest flesh-eating 'fish'. We humans first cook our food, and then eat it, but the great white shark eats its food, and *then* cooks it.

Most people are paranoid about sharks. They think of the dorsal fin cruising through the water, the cruel ramming snout, those endless rows of teeth, and the newspaper headlines. They think of the shark as just a mindless eating-machine, constantly eating, and constantly turning fish, seal or tuna into more shark.

But it is a long time between feeds for the great white shark. Luckily it has a huge liver, about 20 per cent of its body weight. By comparison, the human liver is only about 5 per cent of our body weight. The shark can live off the fat of that liver for a few weeks. In the world of the great white shark, food comes at unpredictable and irregular intervals. The great white shark has to make the most of the food while it is there. Wouldn't it be terrible, if it just had a big meal of delicious baby seal, and its tummy was full, and then along came another baby seal, and it could not eat it because its tummy was already full.

Luckily, the great white shark has a trick to empty its stomach rapidly. It can sort of internally cook its food by heating up its stomach. A hot stomach means that the digestive juices and enzymes in the stomach work faster, and break down the food faster. And once that meal is out of the way and stored in the liver, it can eat another meal straight away, and then store it in the giant liver as well. The great white shark is the ultimate fast-food eater.

Sharks are flesh-eating 'fish'. All fish and practically all sharks are cold blooded, except for four sharks that can raise their body temperature above the water temperature. The great white shark is one of them. This is how it does it. The blood that comes out of exercising muscles is hot. Now fish and most sharks simply dump this heat into the surrounding water, in the same way that the standard radiator in your car dumps excess heat into the airstream. But the great white shark has a clever arrangement of blood vessels that feed this hot blood to the stomach

DON'T
BOTHER
HEATING
IT . . . .

wall. It's a bit like dumping the heat from the hot engine water in your car into a special radiator that is mounted inside the passenger compartment. The stomach can get quite hot. In fact, scientists put a thermometer inside a blue fin tuna, and fed it to a 360-kilogram 3.5-metre male great white shark. They found that the shark could heat its tummy up by 8.3C°. And 24 hours after the feed, the stomach was still 7.4C° hotter than the surrounding water.

All this work was done by a pair of American scientists at a place called Dangerous Reef off South Australia. They found that the great white shark really does have fire in its belly. Sharks have a reputation for being bad tempered, but the great white shark is probably the angriest, because it's simmering inside.

REFERENCES

New Scientist No 1559, 7 May 1987, p 31
Discover February 1988, p 8

Sleep and the Siesta

a Boeing 707 overshoots Los Angeles Airport at a height of 10 kilometres. It keeps on flying out over the Pacific. Its entire crew are fast asleep—why? Their sleep patterns were disrupted, and they were catching up with a little siesta.

The albatross and the shark almost never sleep, while the giraffe gets only 20 minutes a day. At the other end of the scale, lions sleep for about 16 hours a day, cats snooze for about 18 hours a day, while gorillas and opossums just have to have their 19 hours a day. But the dolphin has the most remarkable sleeping pattern of all; it only sleeps with half of its brain at a time, so first the left brain sleeps, and then the right brain, and so on.

But in humans, there are four stages of sleep. In the first light stage of sleep, we are barely under. We could be awakened easily and not even realise we have been asleep. But after a few minutes we slide down to stage two of sleep. Our eyes roll slowly from one side to another, but if someone lifts our eyelids we don't respond. We've been asleep for about 10 minutes by now, but if someone wakes us with a quiet cough, we would believe that we had been awake all this time. In the third stage of sleep, it takes quite a loud noise to wake us. After about 25 minutes, we reach stage four of sleep—the deepest level. We don't dream much down in stage four. After about 20 minutes in stage four we begin to drift up, back into the lighter levels.

Ninety minutes after we fell asleep, we enter a special type of stage one sleep called REM or Rapid Eye Movement sleep. This is where we do our dreaming. Our eyes jerk about under our closed eyelids as if watching a tennis match. We are hard to wake up, but if we are woken up we almost always have vivid memories of colourful dreams. Dolphins don't have REM sleep. After about 10 minutes in the REM sleep, we normally wriggle around in bed, turn over and then dive down towards stage four again. Each night, we go through this cycle about four or five times.

Aristotle thought that we slept because warm vapours arose inside the stomach. That was why people became quite sleepy after a big meal. We now know that Aristotle was probably wrong, but we still don't know why we sleep.

But it does seem that in humans the after-

noon nap or siesta is a natural event. People in an agricultural society will usually have a nap around midday, whether they are in Scandinavia or in the tropics. In laboratory experiments, where people were kept away from a daily nine-to-five schedule and coffee, the sleep researchers found that people wanted to sleep in the early afternoon. This siesta research was done by Jurgen Zulley and Scott Campbell at the Max Planck Institute of Psychiatry in Munich. They found that our industrial society has stolen away our natural afternoon nap.

And not only that, once people were used to their midday siesta, they then wanted to have a morning nap at nine o'clock and a late afternoon nap at five o'clock. This is the pattern of three-month-old babies and it's probably the natural pattern of a human adult.

But it was the doctors at the largest hospital in Athens who found that if you had a 30-minute afternoon siesta, you could cut down your chances of heart disease by 30 per cent. That's amazing—an afternoon nap gives you more protection against heart disease, than a cholesterol-free diet or an exercise program. What a shame that the Greek government then passed an act of parliament that outlaws the afternoon siesta in shops.

So it's practically our duty to reduce the health costs for our country by having an afternoon nap. So go out there and sleep for Australia.

REFERENCES

New Scientist No 1547, 12 February 1987, p 33
The Lancet 1 August 1987, pp 269, 270
Biological Psychiatry Vol 22, 1987, pp 931–932
New Scientist No 1594, 7 January 1988, pp 60–62

Spiders' Sticky Webs

*i*n a horror movie one of the worst things that can happen is to run your face into a spider's sticky web. But not all of the spiderweb is sticky, just bits of it. Unfortunately, spiders are so hated and feared by humans, that they had to invent a special disease—Arachnephobia.

But practically all spiders are harmless to humans. In fact, spiders do the human race a great service by catching and eating billions of insects every day. There are at least 30 000 different species of spiders, but only 2500 of them make a web. And it is only in the late 1980s that scientists discovered how the web really works.

People just don't think of spiders as being caring, sharing critters. But even if you're a death-dealing killer spider, with drops of deadly poison dripping off every fang, like the infamous funnel-web, you can still have motherly love. After mummy funnel-web gives birth to the spiderlings, they don't move much during their first week of life and mummy ignores them. But then the spiderlings perk up and lurch over to mummy, and stroke her with their eight little legs. Mummy is deeply touched by this affection, and gives some of her meal to some of her little babies. In fact, sometimes the spiderlings get naughty and try to steal some food from their mother.

Spiders evolved shortly after the first insects appeared, about 370 million years ago. They very rapidly developed fangs and venom glands loaded with some very fancy drugs. Practically all the spider venoms act like nerve gas, scrambling the nervous system of their prey. But as the spiders evolved further, they invented the web.

Some modern tricky spiders have gone high-tech. Not only are their webs thin and sticky, but to an insect, they glow whiter than white, and in fact, they look like a sexy plant. We humans can't see ultraviolet light, but insects can. Many flowers have patterns of ultraviolet reflection, near where the pollen and nectar are. This pattern spells FOOD to insects. Some of the spiders that make long-term webs (the ones that they leave up for days at a time) have added a chemical to their webs to make them reflect ultraviolet light in the same pattern that the flowers use. And it works really well—these glowing spiderwebs capture 60 per cent more insects than the ordinary nonfluorescent types.

But besides catching insects, spiderwebs have other uses. From the eighteenth century on, astronomers and surveyors used them to make crosshairs. Up until the invention of modern synthetics, spiderweb was the strongest and thinnest fibre around. Strands taken from a spiderweb were used to make the calibrating crosshairs in the 26-inch (66-centimetre) refracting telescope at the Royal Greenwich Observatory in England. They would observe when the sun hit the crosshairs, and this would give them midday. From this standard, all the clocks in the world were set. But recently, a fly found its way inside the telescope and snapped the old dried out strands of spiderweb. When the astronomers tried to get a replacement crosshair, they found nobody knew any more how to convince a spider to spin its web in the right place at the right time. They ended up using crosshairs made out of nylon. But nylon is twice as thick as the spiderweb, so the telescope is not as accurate as it used to be.

Now you might think that all the strands in a spiderweb are the same, but there are two different types, stretchy ones, and not-so-stretchy ones. When the garden cross spider *Araneus diadematus* makes its web, it extrudes only one type of thread. But depending on where the thread will be used, it either puts a coating on it, or it doesn't. This coating radically changes the properties of the thread.

First there are the uncoated radial strands, like the spokes of a bicycle, which all radiate out from a common centre. These strands are very stiff, and will only stretch by about 20 per cent before they snap. The other strands are the coated sticky spiral capture strands, which go round and round from the centre like the grooves of an old-fashioned LP record. These spiral capture threads can stretch up to 200 per cent of their original length, and then contract again into a tight spiral without sagging. They're called 'capture threads' because they're made sticky with a thick glue that is 80 per cent water, and 20 per cent amino acids, fats and salts. This glue is extruded at the same time as the thread, from the front end of the spider. It comes out as a cylinder of sticky liquid, but it immediately bunches up on the thread into a series of evenly spaced droplets about 100 microns apart. That spacing is a little bit more than the thickness of a human hair. But as these droplets of glue begin to globulate, they act like tiny winches, and they wind up balls of loose spiderweb thread inside them, and keep the web nice and tight. So when you

try to stretch the threads of the capture spiral, the extra loops of thread inside these droplets of glue unwind—and when you take the load off the thread, they spring back and coil up inside the drops of glue again.

This research was done by two scientists at the University of Oxford—Fritz Vollrath, a zoologist and Donald Edmonds, a physicist. They reckon that having stiff radial threads and elastic spiral threads woven into a single web gives it the ultimate design.

The nonsticky stiff radial threads provide the basic structure to the spiderweb. They let the spider move around on the web without getting stuck, and let the spider know when something is caught in the web by transmitting the vibrations of the prey as it wriggles around.

The sticky glue-coated elastic spiral capture threads can maintain the tension whether they're stretched or contracted. Because the threads are elastic, they tend not to touch each other when the web is pushed around in a strong wind. These elastic properties are very handy for absorbing the energy of a fast-flying insect, for helping the insect entangle itself in the sticky web, and finally, for not giving the struggling insect a solid footing that might help it escape.

The spiders have evolved a brilliant design for their web. And by coating the right parts of their nontangled web with a high-technology glue, they're able to stretch their net beyond the limits of insect technology.

REFERENCES

Science Digest August 1988, p 24
Nature Vol 340, 27 July 1989 'Modulation of the Mechanical Properties of Spider Silk by Coating with Water' by Fritz Vollrath and Donald T. Edmonds, pp 305–307
New Scientist No 1678, 19 August 1989, p 15
Science Digest January 1990, pp 61–62

Whales Till the Ocean Floor

S trange huge shallow trenches, the size of semitrailers, have been discovered scooped out of the mud on the ocean floor, between Siberia and Alaska.

They were discovered by two American marine geologists, Hans Nelson and Kirk Johnson. They discovered the trenches by towing an underwater sonar unit behind a boat. The sonar unit gave out sound waves, just like the ultrasound machine used in pregnancy imaging. But instead of pictures of babies, they got pictures of a ravaged sea floor covered with shallow grooves and trenches. They found that 20 per cent of the ocean floor looks like a battlefield.

What sea creature could create such havoc? Whale hunters of the nineteenth century sometimes saw California gray whales at the surface with plumes of muddy water gushing from their mouths.

Each year 16000 California gray whales migrate 8000 kilometres from Baja, California to the Bering Sea. They travel there to each eat more than a tonne of food every day. They have their little 2-tonne babies with them. The babies have to eat if they want to get bigger, and there's more food in colder water than in warm water. So they breed in the warm water but feed in the cold water.

Once they get there they dive down to the 50-metre deep ocean floor where they graze like horses or cattle. They feed by sucking their food out of the mud. The mud is riddled with a type of prawn called an amphipod. Out of each 6 kilograms of mud, 1 kilogram is pure wriggling delicious and nutritious amphipod.

The whales feed by rolling on one side while swimming very close to the ocean floor. Then they give a big suck with their huge tongue. At the same time, they open their cavernous mouth, on the side closest to the ocean floor and suck in mud (and a little water).

Now these whales have a mesh that hangs from their upper jaw, where teeth would be if they had teeth. The water is forced out through that mesh. Anything bigger than 4 millimetres is trapped, and they gulp it down and polish it off on the next swallow. Anything smaller is flushed out with the water.

In a feeding season (from May to November) 16000 gray whales eat 30 million tonnes of amphipods. They turn amphipod into blubber. And to get at 30 million tonnes

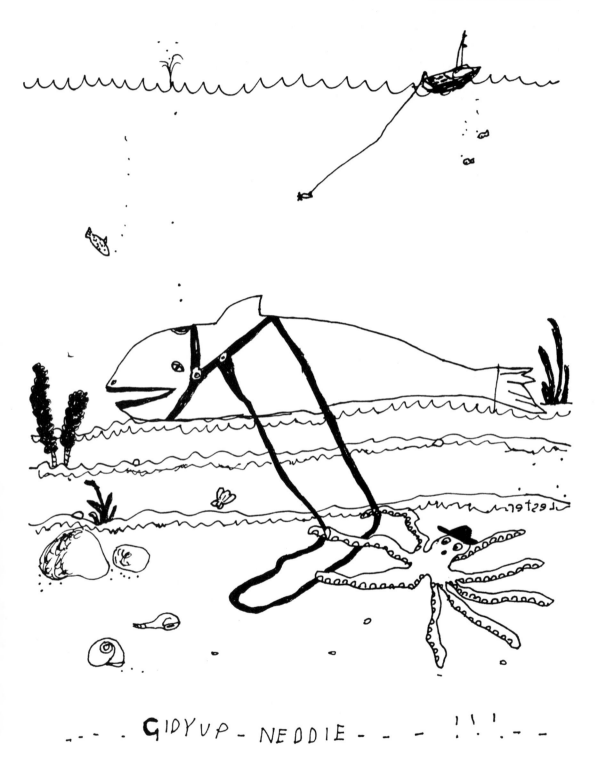

of food, they have to shift 180 million tonnes of mud. That's three times more mud than the mighty Yukon River in Canada dumps on the ocean floor each year.

And then they 'reseed' this freshly disturbed mud with tiny amphipods squirted out through the mesh—so they sow a new crop, just like farmers sow seed into freshly ploughed soil. So next year, there'll be more little amphipods to eat.

They're the original dirt farmers—they sow and they reap. They toil in the soil for spoil.

REFERENCES

Scientific American February 1987 'Whales and Walruses as Tillers of the Sea Floor' by C. Hans Nelson and Kirk R. Johnson, pp 74–81
Scientific American April 1988 'The Behaviour of Baleen Whales' by Bernd Würsig, pp 78–85

Wooden Chinese Spaceships

*t*he Chinese are successfully launching and recovering wooden spaceships.

They actually make most of each spaceship out of various metals, but they do make the heat shield—the part that's exposed to temperatures hotter than the surface of the sun—of each returning spaceship out of wood.

Many countries, including France, Japan, Sweden, India and Israel can now launch spacecraft but only three can get them back again. These three countries have mastered the space-faring techniques needed to have a controlled re-entry and recovery. The returning craft has to be aimed at a very precise angle into the atmosphere. Too steep an angle and it will go too fast and burn up, while too shallow an angle means that it will bounce off the upper atmosphere back into space.

So not only can these countries launch a ship into space, but they can also recover it, in one piece, down here on the ground. And that's important, if you want to send up astronauts who can live to tell the tale. These three countries are the USA (who have a launch about once every two months), the Soviets (who pop up a spacecraft every three days or so), and the Chinese. The Chinese have had a 100 per cent success rate with their dozen recoveries. In fact, in their whole space program, they've had only one failure out of their last 24 launches.

The Chinese have two launch sites. The one at Xichang is for geostationary satellites, which stay in orbit over a fixed place on our planet. They have a separate launch site in Jiuchuan for remote-sensing earth-observation satellites, some of which come back to earth. They are also building a third launch complex at a site south of Beijing. Space is big business for the Chinese. Not only are they very dependable and punctual with their space launches, but they charge less than anyone else. That's why Australia is paying the Chinese to launch the next two Aussat communication satellites in 1991 and 1992.

The Chinese have done lots of research with their recoverable satellites. They've done experiments with algae for the French—algae could be an important source of food on a long space flight. They've found mineral deposits by looking at space photos. Another satellite photograph averted possible disaster in an area prone to earthquakes—

it showed a geological fault zone running through the site for a future electrical power station. They quickly found another location.

The Chinese don't have the sophisticated technology that the Americans have, so they can't yet convert a picture into electronic pulses, send it back to earth, and still maintain super-high resolution. But it's easy to get super-high resolution film, so it makes sense for the Chinese to send back photos.

As a spaceship comes back into the atmosphere, it's travelling at tremendous speed, about 27 000 kilometres per hour. At this enormous speed, there's enough friction between the atmosphere and the skin of the spaceship to melt it into an expensive blob. So a heat shield of some heat-resistant material is used.

On the space shuttle the Americans use a fancy low-density silica ceramic—those famous tiles. In fact, they call the space shuttle 'The Flying Brickyard'. The tiles are very expensive to make, and slow and tedious to glue into position, but they work. But the Chinese use a heat shield of wood—about 15 centimetres of oak!

When the Chinese spaceship hits the atmosphere at a very great speed, it's also at a very high altitude. There's not very much oxygen, so the wooden shield doesn't burn, it just chars into charcoal. Within seconds, the temperature on the heat shield is hotter than the surface of the sun—more than 5000°C!

As the satellite descends, the outer layer of the charcoal gets stripped off by the violent wind, molecule by molecule. At the same time, a bit more wood under the outer layer turns into charcoal. So there's always a layer of charcoal on the surface of the wood, even though the total thickness is always getting less. But charcoal and wood are great insulators, so very little heat gets through to the metal underneath.

The Americans even used cork as a heat shield while launching some vehicles! When they were building an anti-ballistic missile system in the mid-1960s, they tried to shoot down the incoming missiles, travelling at about 5 kilometres per second, with ground-launched anti-missile missiles. They found that for the system to work, the anti-missiles had to take off very rapidly—so rapidly, that the skin temperature of the interceptor anti-missile was higher than the inside of the rocket engine! And so they used cork as a heat shield.

Now NASA did test mahogony, maple and balsa wood for heat shields in the early 1970s. But they found that the layer of char, the burnt wood, is quite weak. The stresses of re-entry such as the pressure, the vibration, and the wind at many kilometres per second, can break it off in large chunks, not molecule by molecule. If big chunks of char break off, you can run out of wood, as your heat shield literally goes up in smoke. No heat shield means a melted spaceship.

But so far, the Chinese have had no problems with the wooden heat shields. The Americans are rapidly pricing themselves out of the space launch market with The Flying Brickyard, while the Chinese are still doing great business with their weatherboard model. So if you're out in the treeless Mongolian desert, and someone yells out 'Timber'—look out for falling spaceships. Don't worry, it's oak-K to run!

REFERENCES

Spaceflight News September 1987, p 6
Time 7 November 1988, p 61
Spaceflight News November 1988, pp 5, 30–33
Spaceflight News October 1988, pp 28, 29

Yawn

*C*ats yawn, babies only five minutes old yawn and even fish and birds go for the big stretch.

You're more likely to yawn just after you've woken up or just before you go to sleep or if you see someone else yawning. But we still don't know why we yawn, or even which parts of the brain control yawning.

When male rats yawn, they have an erection. When Siamese fighting fish are on their own, they hardly ever yawn. But if you put another Siamese fighting fish in the same tank, they will yawn 300 times more often. And once they are actually fighting, their yawning rate doubles again. Charles Darwin noted in 1873 that if a baboon is involved in a threat, or a matter of passion, it is more likely to yawn.

A yawn is a long slow deep inward breath, taken with a wide-open mouth and followed by a short outward breath. But yawn scientists have measured that most yawns are not accompanied by a long slow luxurious stretch of arms and legs—only one in nine. But on the other hand, about half of all stretches will be accompanied by a yawn.

Yawning does have a medical significance. For some unknown reason, schizophrenics hardly ever yawn. On the other hand, people suffering from certain types of epilepsy, encephalitis, damage to the medulla and a few types of brain cancer often can't stop yawning. And for yet another unknown reason, people suffering from an acute physical illness will show that they are beginning to get better by starting to yawn.

Yawning can even move paralysed arms. In a stroke, a small part of your brain dies and this can leave you with a paralysed arm or leg. But there have been a few cases where stroke victims could move their paralysed arms in a stretching motion while they were having a yawn, even though they couldn't move their arms when they wanted to.

But why do average people yawn? One theory was that you would yawn if you had too little oxygen in your blood or too much carbon dioxide. But Robert Provine at the University of Maryland, USA, one of the world's leading yawn experts, did an experiment with first year university students that disproved this theory. He had them breathing air containing different amounts of oxygen and carbon dioxide, and then he counted how often they yawned. The amount of oxygen or carbon dioxide in the air had absolutely no effect on yawning—even when they breathed 100 per cent oxygen, their yawning rate stayed the same.

But yawning has one important effect on the lungs. The alveoli are those tiny air sacs in your lungs where you absorb oxygen and dump carbon dioxide. If you do a lot of quiet and very steady breathing, like when you're asleep or watching boring TV, the alveoli in your lungs start to collapse. A single yawn will open up your alveoli, but so will a deep sigh.

And of course yawning is very important

when you travel on aeroplanes. Yawning opens up your eustachian tubes, which run from your middle ear to the back of your throat. So a few good yawns will relieve the pain of the air pressure when the plane is taking off or landing.

People are most likely to yawn in the first hour after they've woken up, in the last hour before they go to sleep or when they're tired, bored or doing long boring repetitive tasks. People riding on fairly empty trains late in the day tend to yawn a lot.

Ronald Baenninger, a yawn expert at Temple University in Philadelphia, has also done a lot of yawn research. The highest yawning rate he has ever recorded was in a calculus class held by the mathematics department, where each student yawned 25 times per hour.

You could think that a reasonable explanation for why you yawn, is that you are tired and you have to stay awake. But this doesn't explain why people who are highly aroused and tense also yawn—people like students, who yawn an awful lot before their exams; athletes, who yawn before the start of their events; and musicians, who can't stop yawning before they go on stage.

Another big mystery about yawning is, why does seeing another person yawn set you off? In fact, even thinking about yawning, or reading this story, will also trigger yawning. If you smile, you'll probably get a few people to smile back, but yawn, and the whole room yawns with you. And if you don't believe me, ask yourself if you've yawned at least once during this story. And if you haven't, thank you for being so interested, and good night.

REFERENCES

The Lancet 6 February 1988, p 300
The Lancet 12 March 1988, p 596
The Lancet 23 April 1988, p 950
Discover June 1989 'The Big Yawn' by Patrick Huyghe, pp 78–81

Tea Cures Asthma

*i*f you kill those 30 000 strangers that you sleep with every night, you could prevent asthma. You use tea; it's an Australian invention and it's worth over $20 million a year.

Asthma is the number four medical killer in Australia—coming after heart attack, cancer and stroke. It kills more people than AIDS. And the teenage death rate from asthma has actually increased! At some time in their life, one out of every 10 Australians will suffer from asthma.

Asthma is a disease of the smaller airways. The airways are pipes that carry air deep into the lungs. In an asthma attack, the muscles in the airways tighten, the mucous lining swells and mucus is produced and lines the walls, making the airways narrow. The wheezing noise you hear asthmatics make is the air whistling through these narrow pipes. The airways of asthmatics are too sensitive, so they react too much and unnecessarily to things like a change in the weather or exercise or various chemicals found in plastics and paints or the dust of western red cedar timber, and especially chemicals from the house dust mite—and this is where the tea comes in.

The house dust mite is a little critter, actually a spider, with eight legs. It lives off the 600 000 flakes of skin that drop off your body every day—that's about 2 grams. Up to 7000 dust mites live in each square metre of your blankets. They're harmless—all they want to do is eat unwanted and unused human skin. But each day, each dust mite squeezes out 20 tiny droppings. By a terrible coincidence, chemicals that are found in the dust mite and in dust mite droppings will set off eight out of 10 asthmatics.

These chemicals from the dust mite are really tough. They're unaffected by the strongest household cleaners, and even acids and alkalis. But strong tea will break up these chemicals in the dust mite droppings and stop them from setting off asthma attacks. They used this idea in the good old days—the female domestic engineer used to sluice out the floor boards of the family home with cold tea. This cut down the dust, but little did they know that this would attack the house dust mite *and* the asthma.

But what about sucking up the little critter in a vacuum cleaner? To a dust mite, fibres in the carpet are like immense tree trunks. As

the vacuum cleaner approaches, a willy-willy rocks the solid tree trunks and sucks up all the dust mites, as well as their millions of droppings and dried dead ancestors. Because the mites are so light, they survive the roller coaster ride up the nozzle, and they land safely in the vacuum cleaner bag. They have reached the highest possible concentration of human skin flakes in the known universe.

Most of the dust sucked up by the vacuum cleaner hits hard against the back wall of the dust bag. Its paper is made of little fibres which are about 5 microns apart. This mesh will catch the dust mites, which are about 40 microns in diameter (about half as thick as a human hair). But the dust mite droppings are much smaller and are shot out straight through the paper bag, and into the air again. The poor human being pushing the vacuum cleaner gets blasted with millions of dust mite droppings with each step. The standard vacuum cleaner is probably one of the best ways of spreading dust mite droppings, and asthma, all through your house.

So if you can't vacuum them up, how can you get rid of the dust mite droppings? Well you can't, but Wes Green of the University of Sydney has found the active ingredient in tea that actually destroys the chemicals from the dust mite. It's plain old tannin, the stuff that makes tea brown and leather tough. So the tannin in the tea on those ancient floor boards was preventing asthma.

This treatment is now available via your local chemist. They can come and treat your home and kill the dust mites and destroy the dust mite chemicals for a few months. And their special chemical doesn't stain your carpet or sheets like tea would.

But this is only part of treating asthma. You need to understand asthma, the drugs used to treat it, and how to take them. And no smoking—40 per cent of those who died from asthma were smokers.

Dust never sleeps—and it dust mite be that tea is not only good for the digestion, but also the lungs!

REFERENCES

The Lancet 21 July 1984 'The Abolition of Allergens by Tannic Acid' by W. F. Green
University of Sydney News 26 March 1985, p 53
Sydney Morning Herald 17 November 1987, p 1